Published by
Michigan State University
Extension
NCR 45

Diseases of Tree Fruits in the East

By:
Alan L. Jones, Professor
Department of Botany and Plant Pathology
Michigan State University

and

Turner B. Sutton, Professor
Department of Plant Pathology
North Carolina State University

Acknowledgments

Appreciation is expressed to Bill Shane, Southwest Michigan Research and Extension Center, Benton Harbor, Mich., for reviewing the entire manuscript; to David F. Ritchie, North Carolina State University, Raleigh, N.C., for reviewing sections of the manuscript; and to Eldon I. Zehr, Clemson University, Clemson, S.C., for preparing the description on Gilbertella rot. We thank our colleagues, as indicated in the figure captions, for providing 35 mm slides to supplement those provided by the authors. The help of the Photographic Services section of the Instructional Media Center at Michigan State University in making composite slides and in cropping and duplicating 35 mm slides is greatly appreciated. We thank Nora Harrison for typing assistance. We give special thanks to Leslie Johnson, editor; Terry O'Connor, graphic designer; and Ken Fettig, publications manager, Outreach Communications, Michigan State University, for technical editing, designing the page format and guiding the project through the many stages of production.

This publication was partially funded from an EPA IPM Educational Grant to A. L. Jones under EPA Cooperative Agreement CX820822–01–5 managed by the International Apple Institute.

The use of color in the bulletin was made possible in part by contributions from BASF Corporation, Ciba Crop Protection, Cole Grower Service, DowElanco, Elf Atochem North America, Inc., Merck & Co., Inc., the Michigan Apple Research Committee, the Michigan Association of Cherry Producers, the Michigan Peach Sponsors, the Michigan Pear Research Association, the Michigan Plum Advisory Board, the Michigan State Horticultural Society, Rohm and Haas Company and ZENECA Ag Products.

Photos on front cover: Montmorency sour cherries at Amon Orchards north of Traverse City, Mich., and Imperial Gala apple (latter photograph courtesy of Hilltop Nurseries, Hartford, Mich.).

Table of Contents

To the Grower

After the third edition of this publication had been in print for 10 years, it began to lag behind the current literature significantly. For example, Leucostoma canker and Alternaria blotch were described as new diseases of apple in North America. Descriptions of apple scab, sooty blotch, pear scab, European brown rot and several other diseases needed rejuvenation. Also, the publication was out of print, with more than 75,000 copies distributed since November 1971, when it was first published as a Michigan State University Cooperative Extension bulletin.

We had planned to revise the publication by citing recent research and by improving descriptions and color photos. However, one thing led to another, and now two years later we end with a major rewrite, particularly of the stone fruits section. Although the diseases are presented in the same general format and order as before, most sections have been revised and improved.

This bulletin describes diseases of pome and stone fruit crops grown in the eastern United States. It focuses particularly on diseases that occur in Michigan and North Carolina, where the authors have nearly 50 years of combined professional experience with diseases of these crops. Information from other parts of the eastern United States is also presented as appropriate. The coverage of diseases caused by fungi and bacteria is thorough, but only the most important diseases caused by viruses and other pathogen types are discussed.

Previous editions of this bulletin were popular with growers, educators and students as a reference guide and diagnostic aid. Proper recognition of the problem is the first step in successful and efficient control of tree fruit diseases. Growers who can identify fruit diseases are better prepared to select the most effective control measures. Whenever possible, color photographs of symptoms have been included as diagnostic aids.

Each disease has a developmental pattern or cycle it passes through during the year. Control chemicals and methods are generally effective only at certain times or stages in the disease cycle. If the pathogen is not in a susceptible stage of development when control is attempted, results will be disappointing. This bulletin outlines disease cycles with emphasis on the portions of the disease cycle known to be important in developing successful control procedures.

Disease control programs must be adjusted annually according to prevailing environmental conditions to achieve the best results. Interrelationships between the pathogen, the host plant and the environment are discussed in some detail as a guide in deciding when and how control procedures should be modified. Specific chemical control recommendations for most of these diseases can be found in spray schedules published by various states and provinces.

At the end of each section, a few references are listed as a starting point for locating additional information from research studies on that disease. Listed below are a general reference to the field of plant pathology and several references with descriptions of less common disorders of tree fruit crops that are not described here.

General References

Agrios, G. N. 1988. *Plant Pathology,* 3rd ed. San Diego, Calif.: Academic Press, Inc.

Fridlund, P. R., ed. 1989. *Virus and Viruslike Diseases of Pome Fruits and Simulating Noninfectious Disorders.* Extension Special Publication SP0003. Pullman, Wash.: Washington State University Cooperative Extension Service.

Hogmire, H. W., Jr., ed. 1995. *Mid-Atlantic Orchard Monitoring Guide.* Extension Publication NRAES-75. Ithaca, N.Y.: Northeast Regional Agricultural Engineering Service.

Jones, A. L., and H. S. Aldwinckle, eds. 1990. *Compendium of Apple and Pear Diseases.* St. Paul, Minn.: American Phytopathological Society.

Ogawa, J. M., E. I. Zehr, G. W. Bird, D. F. Ritchie, K. Uriu and J. K. Uyemoto, eds. 1995. *Compendium of Stone Fruit Diseases.* St. Paul, Minn.: American Phytopathological Society.

Pome Fruits

APPLE SCAB

S cab, caused by the fungus *Venturia inaequalis* (Cooke) G. Wint., occurs in most areas of the world where apples are grown. It is less severe in semi-arid regions than in cool, humid areas with frequent rainfall. The climate in apple-growing regions east of the Rockies is extremely favorable for scab.

Symptoms

Apple scab occurs on the leaves, petioles, blossoms and fruit. The most striking symptoms occur on the leaves and fruit (*Photo 1*).

Infections usually develop first on the undersides of leaves on fruit spurs, the side exposed when the fruit buds open (*Photo 2*). Once the leaf has unfolded, both sides may be infected.

Conidia are produced abundantly on new lesions; therefore, lesions appear as velvety brown to olive spots that turn black with age (*Photos 1, 3*). At first, the margins (edges) of the lesions are feathery and indefinite, but later they are distinct.

Severe infection can cause extensive defoliation. Trees defoliated 2 or 3 years in a row are weakened and susceptible to low-temperature damage. Return bloom also may be reduced.

Fruit infections resemble leaf infections when young but become brown and corky with age. Early-season infections often occur around the blossom end of the fruit; later, they can occur anywhere on the surface (*Photo 4*). Scab infections result in uneven growth of fruit and cracking of the skin and flesh (*Photo 5*).

Photo 1. Apple scab lesions on fruit and leaves near midseason.

Photo 2. Primary apple scab lesions on the undersides of McIntosh apple leaves.

Photo 3. Sporulation of *Venturia inaequalis* in scab lesions on a McIntosh apple leaf.

Photo 4. Primary scab infection on the blossom end (calyx) of a McIntosh apple.

If infection occurs in late summer or early fall, rough, black, circular lesions may develop on the fruit in storage. These lesions are usually small, varying in size from specks to ¼ inch diameter, and are known as "pinpoint scab."

Disease Cycle

Primary cycle: The fungus overwinters in leaves on the orchard floor

Photo 5. Cracking of McIntosh fruit associated with large scab lesions late in the season.

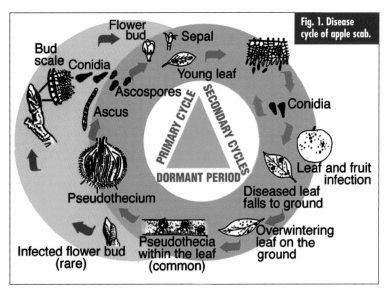

Fig. 1. Disease cycle of apple scab.

(follow Fig. 1 when reading the text on disease cycle) and sometimes in apple buds. In late fall and early spring, microscopic, black, pimple-like structures called pseudothecia are produced in these dead leaves. Within each pseudothecium are asci, each with eight two-celled, olive ascospores (Photo 6). The ascospores produce the first, or primary, infections on the new growth. Conidia produced on the inner surfaces of apple bud scales in orchards that had substantial levels of scab the previous season can also produce primary infections on new growth.

Pseudothecial development is favored by alternating periods of wetness and dryness in late winter and early spring. Normally, some pseudothecia contain mature ascospores when the blossom buds start to open. In some seasons, spores mature by bud break. When leaves on the orchard floor become wet, spores are ejected into the air. Air currents carry them to the emerging tissues, where infection occurs. Maturation and discharge of ascospores usually lasts 5 to 9 weeks.

Germination begins when ascospores or conidia land on young leaves or fruit, provided a film of moisture is present (Photo 7). The number of hours of wetting required for infection varies with prevailing temperatures. For example, at an average temperature of 58 degrees F, primary infection will occur 10 hours after the start of the rain. After 22 hours of wetting, the degree of infection will be severe. Growers can determine whether infection has occurred by noting when a rain begins, when the foliage dries and what was the average temperature during the wet period (Table 1). This simple method is useful in deciding when fungicide sprays are needed.

Lesions are not visible until at least 9 days after the fungus first infects the leaves and fruit. Depending on the average temperature after penetration, 9 to 17 days are required from the onset of infection to the appearance of the olive-green, velvety scab lesions. Secondary spores (conidia), which perpetuate the disease in summer, are produced within these lesions.

Secondary cycle: Secondary infections are initiated by conidia produced in primary lesions. Conidia can be produced as soon as 7 to 9 days after infection, so secondary infection, if not controlled, may be initiated as early as bloom, depending on conditions. Secondary infection at bloom is more common when ascospores infect sepals and leaves at the green tip stage of bud development. Conidia are disseminated by splashing rain and by wind. Conidial germination and infection occur under about the same conditions as germination and infection by ascospores.

Secondary infection on fruit can occur in the fall but not show up until the fruit have been stored for several months. The disease can also build up on the leaves after harvest. Because the fungus overwinters in these leaves, the inoculum level in the spring may be high even though a good spray program was followed the previous year.

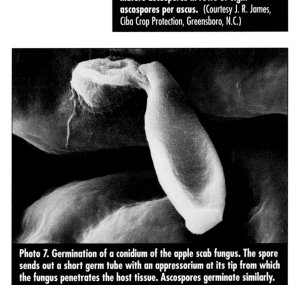

Photo 6. Cross-section of a pseudothecium of the apple scab fungus with pigmented mature ascospores in rows of eight ascospores per ascus. (Courtesy J. R. James, Ciba Crop Protection, Greensboro, N.C.)

Photo 7. Germination of a conidium of the apple scab fungus. The spore sends out a short germ tube with an appressorium at its tip from which the fungus penetrates the host tissue. Ascospores germinate similarly.

Control

Resistant cultivars: Apple breeding programs to develop high quality, disease–resistant cultivars are being carried out at the New York State Experiment Station, Geneva, and cooperatively between Purdue University, Rutgers University and the University of Illinois. Additionally, similar breeding programs exist in a number of foreign countries. More than 25 scab–resistant cultivars have been released, including Prima, Priscilla, Jonafree, Redfree, Liberty, Freedom, Goldrush and Pristine. Most are best adapted to the more northerly apple–growing areas in the United States. The cultivars vary in their susceptibility to other early–season diseases; all are susceptible to summer diseases. Freedom is particularly susceptible to bitter rot. Additional selections are in advanced stages of evaluation.

Sanitation: Prevention of pseudothecial formation in overwintering leaves would probably eliminate scab. Unfortunately, complete elimination of pseudothecia is not possible under orchard conditions with current methods. The potential for severe scab may be reduced, however, by making applications of 5 percent urea to the foliage in autumn to hasten leaf decomposition. Applications should be made just prior to leaf fall to avoid stimulating tree growth and predisposing trees to winter injury.

Chemical control: Apple scab is controlled primarily with fungicide sprays. A variety of fungicides are available; how and when they should be used depends on their mode of action.

Protectant fungicides prevent the spores from germinating or penetrating leaf tissue. To be effective, they must be applied to the surface of susceptible tissue before infection is predicted using the Mills system. These chemicals are applied routinely 7 to 10 days apart, or according to anticipated rainfall (infection periods).

Postinfection fungicides control the scab fungus inside apple leaves and fruit. They can penetrate apple leaves, blossoms and green fruit to inhibit lesion development. The relative ability of fungicides to stop initial infections is referred to as after–infection activity, "back–action" or "kickback" action. The extent of after–infection activity is limited to a few hours or days after the onset of infection, and it often varies with the temperatures that prevail for 24 to 48 hours after spray application.

Other postinfection fungicides can inhibit the fungus even later into the incubation period. Such fungicides have presymptom control activity. The development of chlorotic scab lesions in leaves, or yellow leaves with green circular areas, indicates that the limits of presymptom control activity have been reached (*Photo 8*). Rarely can postinfection fungicides eradicate lesions after sporulation has occurred. Eradication of lesions, referred to as postsymptom control activity, results in lesions that appear to be "burnt out" (*Photo 9*).

Resistance to scab fungicides: Resistance problems with apple scab have occurred with dodine (Cyprex, Syllit) and benzimidazole fungicides (Benlate, Topsin M). Reduced sensitivity to the sterol demethylation inhibitors (DMIs) (Nova, Rubigan) has also been reported but is not widespread in North America. Fungicide resistance may arise

Photo 8. Presymptom scab control. Chlorotic apple scab lesions produced when a sterol demethylation inhibitor was applied after infection but before symptoms were observed.

Photo 9. Postsymptom scab control. Necrotic apple scab lesions produced when a dodine-captan combination was applied to control visible infections.

because of mutation and subsequent selection or through selection of the least sensitive individuals out of a normal population. Resistance to Benlate arises when the scab fungus mutates to produce highly resistant individuals. The resistant strains increase quickly with continued benzimidazole use, and scab control is lost after about 3 years. Reduced sensitivity (a type of resistance) to DMIs arises when the least sensitive members of the scab population survive DMI treatment. These less-sensitive strains, although weakly inhibited by DMIs, increase in the population over time and eventually lead to diminished scab control. Resistance to dodine is similar to the DMI type.

Table 1. Approximate wetting period required for primary apple scab infection at various air temperatures and time required for development of conidia.[a]

Average temperature		Wetting period (hr)[b]			Incubation period[c] (days)
(°F)	(°C)	Light infection	Moderate infection	Heavy infection	
78	25.6	13	17	26	—
77	25.0	11	14	21	—
76	24.4	9.5	12	19	—
63–75	17.2–23	9	12	18	9
62	16.7	9	12	18	10
61	16.1	9	13	20	10
60	15.6	9.5	13	20	11
59	15.0	10	13	21	12
58	14.4	10	14	21	12
57	13.9	10	14	22	13
56	13.3	11	15	22	13
55	12.8	11	16	24	14
54	12.2	11.5	16	24	14
53	11.7	12	17	25	15
52	11.1	12	18	26	15
51	10.6	13	18	27	16
50	10.0	14	19	29	16
49	9.4	14.5	20	30	17
48	8.9	15	20	30	17
47	8.3	15	23	35	—
46	7.8	16	24	37	—
45	7.2	17	26	40	—
44	6.6	19	28	43	—
43	6.1	21	30	47	—
42	5.5	23	33	50	—
41	5.0	26	37	53	—
40	4.4	29	41	56	—
39	3.9	33	45	60	—
38	3.3	37	50	64	—
37	2.7	41	55	68	—
33–36	0.5–2.2	48	72	96	—

[a] Adapted from Mills, 1944; modified by A. L. Jones.
[b] The infection period is considered to start when rain begins.
[c] Approximate number of days required for conidial development after the start of the infection period.

Because of the resistance problems associated with the use of many fungicides, anti–resistance strategies need to be a part of most scab control programs. Today, resistance management is likely to be the most successful with DMIs because they are the most recent fungicides introduced for scab control. The best strategy is to use full rates of DMIs tank mixed with a non–related protectant. Maintaining spray coverage and reasonable spray intervals also help to reduce the number of less–sensitive individuals able to withstand the spray.

Selected References

Becker, C. M., and T. J. Burr. 1994. Discontinuous wetting and survival of conidia of *Venturia inaequalis* on apple leaves. *Phytopathology*, 84:372–378.

Becker, C. M., T. J. Burr and C. A. Smith. 1992. Overwintering of conidia of *Venturia inaequalis* in apple buds in New York orchards. *Plant Dis.*, 76:121–126.

Jones, A. L., S. L. Lillevik, P. D. Fisher and T. C. Stebbins. 1980. A microcomputer-based instrument to predict primary apple scab infection periods. *Plant Dis.*, 64:69–72.

Koenraadt, H., S. C. Somerville and A. L. Jones. 1992. Characterization of mutations in the beta-tubulin gene of benomyl-resistant field strains of *Venturia inaequalis*. *Phytopathology*, 82:1348–1354.

MacHardy, W. E. 1996. *Apple Scab: Biology, Epidemiology and Management*. St. Paul, Minn.: American Phytopathological Society.

MacHardy, W. E., and D. M. Gadoury. 1989. A revision of Mills' criteria for predicting apple scab infection periods. *Phytopathology*, 79:304–310.

Schwabe, W. F. S., A. L. Jones and J. P. Jonker. 1984. Changes in the susceptibility of developing apple fruit to *Venturia inaequalis*. *Phytopathology*, 74:118–121.

Sutton, T. B., A. L. Jones and L. A. Nelson. 1976. Factors affecting dispersal of conidia of the apple scab fungus. *Phytopathology*, 66:1313–1317.

Tomerlin, J. R., and A. L. Jones. 1983. Effect of temperature and relative humidity on the latent period of *Venturia inaequalis* in apple leaves. *Phytopathology*, 73:51–54.

Wilcox, W. F., D. I. Wasson and J. Kovach. 1992. Development and evaluation of an integrated, reduced-spray program using sterol demethylation inhibitor fungicides for control of primary apple scab. *Plant Dis.*, 76:669–677.

Pome Fruits

FIRE BLIGHT

Fire blight, caused by the bacterium *Erwinia amylovora* (Burrill) Winslow et al., has caused significant losses to the apple industry in the eastern United States in recent years. The disease will continue to threaten the industry because of increased planting of commercially valuable but highly susceptible rootstocks and cultivars and the recent development in some areas of the East of streptomycin-resistant strains of the pathogen. Fire blight has largely eliminated commercial pear production in the eastern United States.

Symptoms

The fire blight pathogen kills the fruit-bearing spurs and branches and often entire trees (*Photo 10*). Recognizing the presence of ooze, droplets of a milky to reddish brown, sticky liquid that seep from the surface of infected tissues, is important for distinguishing fire blight from other disorders.

Infected blossoms become water-soaked and darker green as bacteria invade new tissues (*Photo 11*). Within 4 or 5 days, fruiting spurs may begin to collapse, turning dark brown to black on pear and brown to dark brown on apple (*Photo 12*). Temperature, as measured by the accumulation of degree-days (about 103 DD, base 55 degrees F, or 57 DD, base 12.7 degrees C, from infection date), governs the timing of symptom development.

Infected shoots turn brown to black from the tip and bend near the tip to resemble a shepherd's crook (*Photo 13*). When shoots are invaded from the base, the basal

Photo 10. Apple tree with dieback of spurs and shoots caused by fire blight.

Photo 12. Fruit spurs on pear with fire blight–infected blossoms.

Photo 11. First symptoms of blossom blight. Note ooze droplets and discoloration of the blossom on the lower right. (Courtesy W. G. Bonn, Agriculture Canada, Research Center, Harrow, Ontario)

Photo 13. Apple shoot with fire blight. Note crook at tip.

Photo 14. Bacterial ooze exuding from an apple infected with fire blight.

leaves and stem turn brown to black. Leaves may be infected through the petioles, resulting in discoloration of the midvein, followed shortly by a darkening of the lateral veins and surrounding tissues.

Bark on infected branches and scaffold limbs is darker than normal. When the outer bark is peeled away, the inner tissues are water-soaked with reddish streaks when first invaded; later the tissues are brown. The presence of reddening in the lesion helps to distinguish fire blight cankers from other kinds of cankers. As lesion expansion slows down, the margins become sunken and sometimes cracked, forming a canker.

Apple and pear fruit infected with fire blight develop a brown to black decay. During wet, humid weather, infected fruit often exude a milky, sticky liquid, or ooze, containing fire blight bacteria (*Photo 14*).

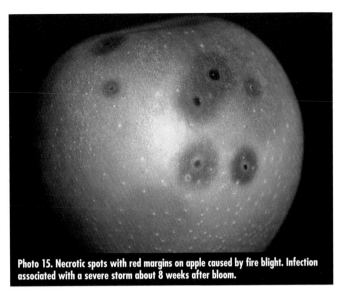

Photo 15. Necrotic spots with red margins on apple caused by fire blight. Infection associated with a severe storm about 8 weeks after bloom.

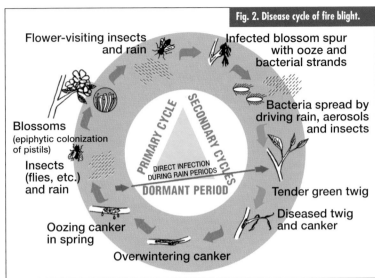

Fig. 2. Disease cycle of fire blight.

Flower-visiting insects and rain
Infected blossom spur with ooze and bacterial strands
Bacteria spread by driving rain, aerosols and insects
Blossoms (epiphytic colonization of pistils)
Insects (flies, etc.) and rain
Tender green twig
Diseased twig and canker
Oozing canker in spring
Overwintering canker
PRIMARY CYCLE
SECONDARY CYCLES
DIRECT INFECTION DURING RAIN PERIODS
DORMANT PERIOD

Photo 16. Decline of Spy apple trees due to infection of the Mark rootstock by fire blight.

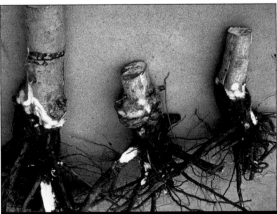

Photo 17. Fire blight on apple rootstocks just below ground level. Bark cut away to show the margin between healthy and diseased crown and root tissues. (Courtesy S. V. Thomson, Utah State University, Logan)

After severe storms, small, black, raised blisters with large, red margins develop on fruit surfaces (*Photo 15*). These fruit develop a firm rot and eventually shrivel into mummies that may persist on the tree.

Trees with roots killed by fire blight show typical symptoms of tree decline in the mid- to late summer (*Photo 16*). Infection of susceptible M.9, M.26 and other apple rootstocks can occur by spread of the bacterial cells through internal tissues of symptomless scion down trunks into the rootstock (*Photo 17*). Infection of the rootstock can also occur from infected suckers arising from rootstocks.

Disease Cycle

The pathogen overwinters near the margins of cankers (*follow Fig. 2 when reading text*). Survival of bacteria is most likely in cankers with indefinite margins following mild winters. Ooze begins to appear on the surface of the cankers at or just before the onset of bloom in the spring.

Before primary infection can occur, the bacteria must move from overwintering cankers to the flowers. This occurs through the action of splashing rain and, occasionally, flies and other insects that visit both bacterial ooze and blossoms. Only a small portion of the blossoms are inoculated in this manner, however. Eventually, honeybees visit infected blossoms and pick up pollen or nectar contaminated with bacteria. Spread of bacteria from flower to flower by bees is rapid. The bacteria can colonize the stigmatic surface of the pistils of healthy apple and pear flowers to produce high epiphytic populations of bacteria.

Climatic conditions are important in determining amount of spread and severity of blossom blight. Temperatures between 65 and 86 degrees F, accompanied by rain or high humidity during the day, favor infection. Although bacteria invade the flowers primarily through natural openings, storms containing wind-driven rain or hail are important in spreading bacteria during the summer, leading to sudden and severe outbreaks of disease. Inoculum for secondary infection originates from droplets of ooze produced on infected flowers, fruit and shoots.

Control

Many practices help reduce the severity of fire blight. Not all measures are necessary or feasible in every planting. No single control method is adequate, however, and a conscious effort must be made to control the disease each year.

Removing sources of infection: In summer, inspect orchards weekly beginning about 10 days after petal fall to remove overwintering cankers and infected spurs and terminals. If infections are cut out rather than broken out by hand, sterilize the cutting tools between trees with sodium hypochlorite (household bleach–Clorox and other brands) diluted 1:10 with water. Cut infected branches far below (at least 12 to 18 inches) the lowest evidence of disease to ensure elimination of the infection. Dormant pruning to remove overwintering cankers is also very important. Make cuts about 4 inches below any evidence of dead bark killed by the disease.

Insect control: Sucking insects such as aphids, leaf-hoppers, plant bugs and pear psylla create wounds that bacteria can enter and at times may spread bacteria.

Resistant cultivars: When establishing new orchards, consider cultivar susceptibility. Although apple cultivars vary in resistance to fire blight, none are immune. Some cultivar/rootstock combinations are too susceptible to fire blight to be grown successfully in parts of the eastern United States. Plans for converting old orchards to high density systems and new cultivars need to include a realistic plan for controlling fire blight without streptomycin.

Susceptible apple cultivars include Braeburn, Gala, Fuji, most strains of Jonathan, Rome, Ida Red, Ginger Gold, Nittany, Red Yorking, Earligold, Twenty Ounce, Rhode Island Greening, Yellow Transparent, Puritan, Wealthy, Lodi, Fenton (Beacon), Mutsu (Crispin), Granny Smith and many crabapple cultivars. Some years, Golden Delicious, Delicious, McIntosh and Stayman develop spur infections, but the disease seldom enters the main branches. M.26 rootstocks are extremely susceptible to fire blight, so fire blight–susceptible cultivars should not be planted on them. Bartlett, Bosc and Clapp's Favorite pear cultivars are highly susceptible. The USDA has a pear breeding program to develop blight-resistant cultivars; however, those cultivars released (Magness, Moonglow and Potomac) are mainly used for backyard production. Additional blight-tolerant cultivars (Harrow Sweet, Harrow Delight) were released from the Harrow Station in Canada.

Cultural practices: Lesion development and damage from blight is more severe when tree growth continues late into the season. Use management systems that promote early cessation of growth without unduly reducing tree vigor. Plant orchards on well drained soil, apply nitrogen fertilizer early and, to stop growth, avoid cultivating orchards later than midsummer.

Chemical and biological control: Sprays of copper applied at ¼-inch green tip may reduce the amount of inoculum on the outer surfaces of the trees. Antibiotics have been highly effective against the blossom phase of fire blight. These sprays are important because they often prevent a fire blight problem from getting started in an orchard. Predictive models, particularly MARYBLYT, help growers identify potential infection periods. Such information is helpful in timing antibiotic treatments and for avoiding unnecessary treatment. Strains of *E. amylovora* resistant to streptomycin are present in some orchards in the eastern United States and in most apple and pear regions in the western United States. Biological control agents have provided partial control and, although not widely used today, may be used in the near future.

Selected References

Chiou, C.-S., and A. L. Jones. 1995. Molecular analysis of high-level streptomycin resistance in *Erwinia amylovora*. *Phytopathology*, 85:324–328.

Lightner, G., and P. W. Steiner. 1990. Computerization of a blossom blight prediction model. *Acta Hortic.*, 273:159–162.

Johnson, K. B., V. O. Stockwell, D. M. Burgett, D. Sugar and J. E. Loper. 1993. Dispersal of *Erwinia amylovora* and *Pseudomonas fluorescens* by honeybees from hives to apple and pear blossoms. *Phytopathology*, 83:478–484.

McManus, P. S., and A. L. Jones. 1994. Epidemiology and genetic analysis of streptomycin-resistant *Erwinia amylovora* from Michigan and evaluation of oxytetracycline for control. *Phytopathology*, 84:627–633.

Thomas, T. M., and A. L. Jones. 1992. Severity of fire blight on apple cultivars and strains in Michigan. *Plant Dis.*, 76:1049–1052.

Thomson, S. V. 1986. The role of the stigma in fire blight infections. *Phytopathology*, 76:476–482.

van der Zwet, T., and H. L. Keil. 1979. *Fire blight— A Bacterial Disease of Rosaceous Plants.* Handbook 510. Washington, D.C.: U.S. Department of Agriculture.

van der Zwet, T., and S. V. Beer. 1995. *Fire Blight—Its Nature, Prevention and Control.* Bull. 631. Washington, D.C.: U.S. Department of Agriculture.

Pome Fruits

Photo 18. Necrotic spots with orange margins on apple leaves caused by cedar-apple rust.

Photo 19. Micrograph of a pycnium of cedar-apple rust fungus with a few periphyses extended and some pycniospores. Nectar drops were removed during preparation.

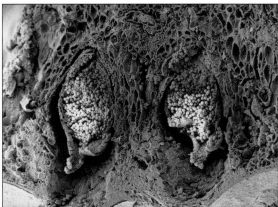

Photo 20. Cross-section of an apple leaf showing young aecia "cluster-cups" of the cedar-apple rust fungus on the underside of the leaf.

RUST DISEASES

Rust diseases are an important economic problem from the Hudson Valley of New York through the mid-Atlantic and southeastern states and from southern Illinois and Indiana through the southern North Central region. They are minor problems from western New York state through Michigan.

The fungi that cause these diseases are cedar-apple rust, *Gymnosporangium juniperi-virginianae* Schwein.; quince rust, *G. clavipes* (Cooke & Peck) Cooke & Peck in Peck; and hawthorn rust, *G. globosum* (Farl.) Farl. All three fungi spend part of their life cycle on the eastern red cedar and are problems only when red cedar is found close to the orchard. The life cycles and control of these diseases are similar. Cedar-apple rust is the most important.

Symptoms

On apple: The bright color of the lesions produced by cedar-apple rust makes it easy to identify (*Photo 18*). Small, pale yellow spots develop on the upper leaf surfaces shortly after bloom. These spots gradually enlarge and turn orange. Orange-colored drops of liquid may be observed in the spots when they are about ⅛ inch in diameter. Later, black dots (pycnia) appear in the spots on the upper surface (*Photo 19*). In late summer, cylindrical tubes or protuberances (aecia) become evident on the leaf undersurfaces (*Photo 20*). The ends of the tubes split open and curl back.

Severe infection results in extensive defoliation and weakens the tree. Fruit infection is most common near the calyx end. Lesions are similar in color to those on leaves except the borders are darker green. Lesions are shallow, not over 1/16 inch deep, and with no internal chlorosis or necrosis. The black pycnia often develop in the lesions, but formation of the cylindrical tubes is less common.

Hawthorn rust produces lesions on apple leaves but rarely on apple fruit. Lesions resemble those of cedar-apple rust but are usually not over ¼ inch in diameter.

Quince rust usually does not cause leaf lesions. Lesions on fruit are dark green, about ¾ to 1½ inches in diameter, and they usually distort the fruit (*Photo 21*). Lesions are deep, with necrotic tissues extending to the core. Affected fruit usually drop before harvest.

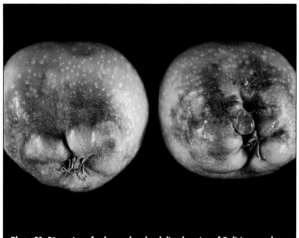

Photo 21. Distortion of calyx end and red discoloration of Delicious apple caused by quince rust.

Quince rust also occurs on pear fruit.

On cedar: *Gymnosporangium juniperi-virginianae* and *G. globosum* produce brown to reddish brown leaf galls from ¼ to 2 inches in diameter. During rainy periods in spring, bright orange, gelatinous spore horns protrude from the galls (*Photos 22, 23*). *Gymnosporangium juniperi-virginianae* galls produce spores for 1 year; *G. globosum* may produce spores for 3 to 5 years. *Gymnosporangium clavipes* produces spindle-shaped swellings (cankers) on branches of cedar. These swellings may be up to 2 feet long and produce spores for many years (*Photo 24*). Cedar trees vary widely in their susceptibility to the rust fungi.

Disease Cycle

The disease cycle of cedar-apple rust is complex (*Fig. 3*). Two plants (apple and cedar) and three fruiting structures (telia, aecia and pycnia) are involved. The pathogen requires two years to complete its life cycle.

The fungus overwinters in reddish brown galls or "cedar apples" in the cedar tree. When wet in spring, the galls extrude gelatinous tendrils or "horns" consisting of two-celled teliospores (*Photo 25*). Each spore produces four basidiospores from each of the two cells. Air currents carry the basidiospores to the apple leaf and fruit. Temperature and wetting conditions favoring infection are very similar to those of apple scab except no infection occurs below 43 degrees F because such temperatures are too cool for basidiospore production. Fruit are most susceptible for a 2- to 3-week period beginning at bloom; foliar infections can occur as long as basidiospores are produced

Photo 22. Tiny cedar-apple rust galls with gelatinous spore horns in axils of cedar leaves.

Photo 23. Cedar-apple rust gall on cedar with gelatinous spore horns fully extended.

Photo 24. Quince rust canker on cedar with gelatinous spore horns fully extended.

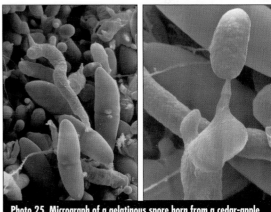
Photo 25. Micrograph of a gelatinous spore horn from a cedar-apple rust gall. Teliospores of the cedar-apple rust fungus with slight constriction at the crosswall and single teliospore bearing promycelium (left) and formation of secondary basidiospore on a pointed sterigmata (right). Other basidiospores can be seen on the left.

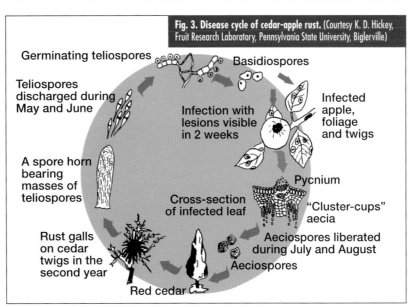
Fig. 3. Disease cycle of cedar-apple rust. (Courtesy K. D. Hickey, Fruit Research Laboratory, Pennsylvania State University, Biglerville)

and new leaves emerge. Leaves are most susceptible when 4 to 8 days old. In July and August, windborne aeciospores from apple infect cedar leaves. Rust lesions develop in 1 to 3 weeks. These galls mature the following summer and produce telia the next spring, approximately 18 months after infection, completing the life cycle of the fungus.

Control

Removing cedars located within a 2-mile radius of the orchard interrupts the life cycle and makes fungicidal control easier. For complete control, remove all cedars within 4 to 5 miles of the orchard.

Fungicides effective against the rust diseases should be applied periodically from the pink stage of bud development to third cover to protect the emerging leaves and developing fruit.

Selected References

Aldwinckle, H. S., R. C. Pearson and R. C. Seem. 1980. Infection periods of *Gymnosporangium juniperi-virginianae* on apple. *Phytopathology,* 70:1070–1073.

Miller, P. R. 1932. Pathogenicity of three red cedar rusts that occur on apple. *Phytopathology,* 22:723–740.

Miller, P. R. 1939. Pathogenicity, symptoms and causative fungi of three apple rusts compared. *Phytopathology,* 29:801–811.

Warner, J. 1986. Susceptibility of apple scab-resistant cultivars to *Gymnosporangium juniperi-virginianae, G. clavipes* and *Botryosphaeria obtusa. Can. Plant Dis. Surv.,* 66:27–30.

Pome Fruits

POWDERY MILDEW

Powdery mildew, caused by the fungus *Podosphaera leucotricha* (Ell. & Ev.) E.S. Salmon, is often an important problem in apple orchards. Losses result from death of vegetative shoots, death of flower buds (and the resulting yield reductions) and loss of fruit quality due to russeting. Powdery mildew usually is destructive only on highly susceptible cultivars, but after it has built up to high levels, it may affect adjacent moderately resistant ones. The problem is more severe when a mildewcide is not included regularly in the scab program.

Symptoms

Mildew occurs in the nursery on terminal growth and in the orchard on leaves, flowers, shoots and fruit. On leaves, lesions first appear as whitish, felt–like patches of fungal mycelium and spores on the undersides and along the margins. The lesions spread rapidly and may engulf the entire leaf. Infected leaves are narrower than normal, are folded longitudinally, and become stiff and brittle with age (*Photo 26*).

Infected blossom buds open several days later than normal or are killed outright by low winter temperatures. When they emerge, flower parts and leaves are usually covered with the white mycelium (*Photo 27*). Infected terminal or shoot buds also produce diseased leaves and shoots. Shoots from infected buds are shorter than those from healthy buds.

Fruit are rarely infected, except on Jonathan, unless the disease

Photo 26. Powdery mildew–infected apple shoot. Note distorted leaves and white growth on leaves and along the shoot.

Photo 27. White growth of powdery mildew fungus on apple leaves emerging from overwintered infected buds.

has built up to high levels. Infected fruit are russeted and sometimes stunted (*Photo 28*).

Disease Cycle

The mildew fungus overwinters as mycelium in buds infected the previous summer (*Fig. 4*). As the infected buds open in spring, the fungus, already established on the emerging leaves, begins to produce spores (conidia). The conidia are carried by wind to the emerging tissues, initiating secondary infections.

Unlike apple scab, which requires a film of moisture to germinate, mildew spores germinate readily in the absence of free moisture at high relative humidities and temperatures between 60 and 80 degrees F. Spore germination and mycelial development are slowed by temperatures between 40 and 50 degrees F, and both are signifi-

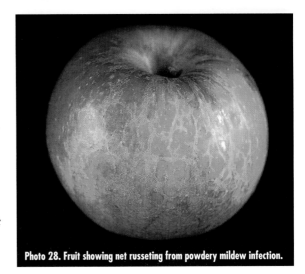

Photo 28. Fruit showing net russeting from powdery mildew infection.

Fig. 4. Disease cycle of powdery mildew.

Infection
Conidia
Haustoria (roots)
Infected leaves and shoots
Infected bud
Cleistothecium
Mildew mycelium
Infected dormant terminal
Infected blossoms and leaves
Conidia
Mycelium
PRIMARY CYCLE
SECONDARY CYCLES
DORMANT PERIOD

cantly reduced at temperatures above 90 degrees F. Mycelium from the germinating spores branches and spreads over the surface of the leaf, putting down small "roots," termed haustoria, into the epidermal cells for nutrition. More spores are quickly produced and the cycle is repeated. Secondary cycles continue to occur until tree growth stops in late summer. Small, dark brown, globular fruiting bodies known as cleistothecia are sometimes formed in the mycelial mat on stems and petioles in late summer. Each cleistothecium has one ascus containing eight ascospores, but it is doubtful they play an important role in overwintering the fungus.

Mildew-infected buds are more susceptible to freezing than healthy buds. At −15 degrees F or colder, most infected buds are killed. Because the mildew fungus is an obligate parasite, it cannot survive after the buds are killed.

Control

Mildew-susceptible cultivars include Jonathan, Rome Beauty, Cortland, Baldwin, Monroe, Ida Red, Granny Smith and Stayman. Where susceptible cultivars are grown, include a mildewcide in the scab program to provide control of both disease. Begin sprays at tight cluster and continue until terminal growth stops. Early sprays (tight cluster to petal fall) are essential to success in controlling powdery mildew.

Selected References

Berkett, L. P., K. D. Hickey and H. Cole, Jr. 1988. Relation of application timing to efficacy of triadimefon in controlling apple powdery mildew. *Plant Dis.,* 72:310–313.

Spotts, R. A., R. P. Covey and P. M. Chen. 1981. Effect of low temperature on survival of apple buds infected with the powdery mildew fungus. *HortScience,* 16:781–783.

Sutton, T. B., and A. L. Jones. 1979. Analysis of factors affecting dispersal of *Podosphaera leucotricha* conidia. *Phytopathology,* 69:380–383.

Yoder, K. S., and K. D. Hickey. 1983. Control of apple powdery mildew in the mid-Atlantic region. *Plant Dis.,* 67:245–248.

Pome Fruits

BLACK ROT

Black rot of apple fruit is caused by the fungus *Botryosphaeria obtusa* (Schwein.) Shoemaker (previously named *Physalospora obtusa* [Schwein.] Shoemaker). *B. obtusa* also causes a leaf spot called frog–eye leaf spot, which is named after the appearance of the lesion. In addition, the fungus causes a limb canker. The limb canker phase is most important in the northeastern and north central apple–growing regions of the United States, and the leaf spot and fruit rot phase are most important in the Southeast.

Symptoms

Leaf symptoms first appear 1 to 3 weeks after petal fall. Leaf infections begin as small, purple flecks that rapidly enlarge to ⅛ to ¼ inch in diameter. The margins of the lesions remain purple and the centers become tan to brown, giving the lesions a "frog–eye" appearance. Heavily infected leaves become chlorotic and drop from the tree (abscise).

Sepal infection may occur prior to bloom, resulting in early fruit abortion or blossom–end rot later in the season. After petal fall, infections on young fruit begin as reddish flecks, which develop into purple pimples. These enlarge into dark brown necrotic areas as the fruit mature. New infections on more mature fruit are often black, irregularly shaped and surrounded by a red halo (*Photo 29*). As lesions enlarge, they are often characterized by a series of concentric rings alternating from black to brown (*Photo 30*). Lesions remain firm and are not sunken.

Pycnidia are usually scattered over the surface of the lesion. Infected fruit mummify and often remain attached to the tree.

Infected areas on limbs and branches are reddish brown and slightly sunken. Some cankers remain small; others enlarge to become several feet long. Infected branches are sometimes weakened enough to break with heavy crop loads; sometimes they are killed outright.

Pycnidia are produced abundantly on limb cankers, in dead twigs, on fruit and occasionally in leaf spots. Pseudothecia are often found in dead wood and cankers.

Disease Cycle

Botryosphaeria obtusa overwinters in dead bark, twigs, cankers and mummified fruit. Ascospores and conidia are released during rainfall throughout the growing season and are washed or blown onto fruit and foliage. Ascospores are generally more common during the spring than in the summer. Sepal infection can occur anytime after bud break, and fruit infection can occur anytime during the growing season. Leaf infections are most common shortly after petal fall. The optimum temperature for leaf infection is 80 degrees F; at this temperature, 4.5 and 13 hours of leaf wetting are necessary for light and severe infection, respectively. More than 24 hours of wetting are required for infection to occur at 50 degrees F or less. The optimum temperature for fruit infection ranges from 68 to 75 degrees F;

approximately 9 hours of wetting are required at these temperatures for infection to occur. Black rot infections commonly develop in leaves and fruit below black rot mummies or old fire blight cankers from conidia from these sources. Early–season infection may result in fruit drop. Severely diseased fruit may mummify and remain attached to the tree.

Photo 29. Initial stages of black rot. Note necrotic spots developing below mummified fruit.

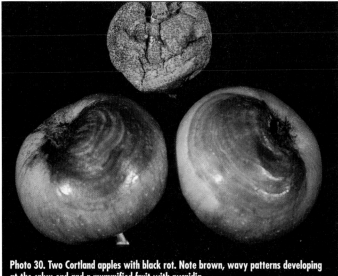

Photo 30. Two Cortland apples with black rot. Note brown, wavy patterns developing at the calyx end and a mummified fruit with pycnidia.

Control

Removing dead wood, mummies and cankers from the trees is important to help control the disease. Current-season prunings should be removed from the orchard or chopped with a flail mower. Prunings piled on the orchard's perimeter can serve as an important inoculum source. Fungicides, applied from silver tip until harvest, are required to control the disease when it is a problem. Cultivars generally do not vary greatly in their susceptibility to *B. obtusa;* however, Empire and Cortland are somewhat more susceptible.

Selected References

Arauz, L. F., and T. B. Sutton. 1989. Temperature and wetness duration requirement for apple infection by *Botryosphaeria obtusa. Phytopathology,* 79:440–444.

Foster, H. H. 1937. Studies of the pathogenicity of *Physalospora obtusa. Phytopathology,* 27:802–827.

Smith, M. B., and F. F. Hendrix, Jr. 1984. Primary infection of apple buds by *Botryosphaeria obtusa. Plant Dis.,* 68:707–709.

Sutton, T. B. 1981. Production and dispersal of ascospores and conidia by *Physalospora obtusa* and *Botryosphaeria dothidea* in apple orchards. *Phytopathology,* 71:584–589.

Pome Fruits

WHITE ROT

White rot, caused by the fungus *Botryosphaeria dothidea* (Schwein.) Shoemaker, is also referred to as bot rot or Botryo-sphaeria rot. The *Fusicoccum* (previously named *Dothiorella*), or asexual stage of *B. dothidea,* is the most common stage found in orchards on infected fruit, cankers or dead wood.

The disease is most common in the southeastern apple–growing regions of the United States, where losses in some orchards have exceeded 50 percent.

Symptoms

Fruit lesions begin as small, often circular, slightly sunken, tan spots, which may be surrounded by red halos (*Photos 31, 32*). On cultivars with red skin pigment, the halo may appear purple or black. As lesions expand, the rotten area extends inward toward the core (*Photo 33*), forming a cylinder of rotten flesh. In more advanced stages, the core becomes rotten and the rotten area advances from the core region into the flesh until the entire fruit is rotten. Under warm conditions (80 degrees F or higher), rotten areas are usually soft, watery and white to tan. If rotten areas develop under cooler conditions, they are usually firmer and a deeper tan and resemble black rot. Scattered clumps of tiny, dark pycnidia develop on the surface of affected fruit. Rotten fruit usually drop, but a few may shrivel and remain attached to the tree.

Photo 31. Early stages of white or bitter rot on fruit.

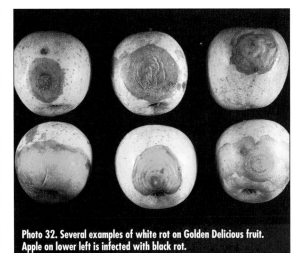

Photo 32. Several examples of white rot on Golden Delicious fruit. Apple on lower left is infected with black rot.

Photo 33. Internal fruit symptoms of bitter rot (left) and apple white rot (right). Decay from white rot (but not bitter rot) usually reaches and surrounds the core.

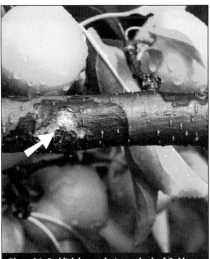

Photo 34. Reddish brown lesion on bark of Golden Delicious apple is the early symptom of white rot caused by *Botryosphaeria dothidea*.

Branch and twig infections begin as a discoloration of the lenticels (*Photo 34*), which later develop into blisters that may exude a watery liquid (*Photo 35*). Branch and twig infections are most severe in hot, dry summers when trees are under drought stress. In older cankers, the outer bark becomes tan to orange and papery, and the margins of the cankers crack and fissure. Large limbs may be killed by fusion of cankers. Pycnidia and pseudothecia are produced in the cankered area.

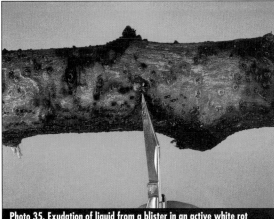

Photo 35. Exudation of liquid from a blister in an active white rot canker that is much older than that shown in Photo 34.
(Courtesy D. A. Rosenberger, Hudson Valley Laboratory, Cornell University, Highland, New York)

Disease Cycle

Botryosphaeria dothidea overwinters in dead bark, twigs and cankers within the tree. Ascospores and conidia are produced throughout the growing season and are washed or blown onto fruit or twigs during rainy periods. Fire-blighted branches and spurs and other dead wood are rapidly colonized and can be an important source of inoculum during the growing season, especially in the warmer growing areas of the eastern United States.

Fruit infection can occur throughout the growing season, but rot symptoms usually do not appear before soluble solids near 10 percent. Fruit infection can occur in as few as 2 to 4 hours at 80 degrees F. Infection of branches occurs through wounds or lenticels. Bark injured by temperature stress is an important infection site. Weak trees exhibiting poor growth and trees under drought stress are often more susceptible to bark infection.

Control

Because *Botryosphaeria* survives in dead wood, pruning and removing all dead spurs, twigs and branches from the orchard is essential. Removing current-season fire-blighted twigs is also important because they can be rapidly colonized by the white rot fungus. Cultivars do not vary greatly in their susceptibility to white rot, though Golden Delicious, Empire and Jerseymac appear more severely affected than others.

A fungicide spray program from bloom until harvest is important to assure white rot control. The dithiocarbamate fungicides are not particularly effective. Trees should be irrigated during hot, dry weather to minimize drought stress and the likelihood of twig and branch infections.

Selected References

Kohn, F. C., Jr., and F. F. Hendrix. 1982. Temperature, free moisture and inoculum concentration effects on the incidence and development of white rot of apple. *Phytopathology,* 72:313–316.

Kohn, F. C., Jr., and F. F. Hendrix. 1983. Influence of sugar content and pH on development of white rot on apples. *Plant Dis.,* 67:410–412.

Parker, K. C., and T. B. Sutton. 1993. Effect of temperature and wetness duration on apple fruit infection and eradicant activity of fungicides against *Botryosphaeria dothidea. Plant Dis.,* 77:181–185.

Pome Fruits

BITTER ROT

Bitter rot is the common name of the fruit rot disease caused by *Colletotrichum gloeosporioides* (Penz.) Penz. & Sacc. in Penz. (sexual stage *Glomerella cingulata* [Stoneman] Spauld. & H. Schrenk) and *C. acutatum* J. H. Simmonds. The fungi also cause a leaf spot and canker. The disease is most common in the warmer apple-growing regions in the eastern United States. However, in 1995 an outbreak of bitter rot occurred in several apple orchards in southwestern Michigan and from Grand Rapids to East Lansing in central Michigan. Of the three rot diseases–bitter rot, white rot and black rot–bitter rot has the potential to be the most destructive in warmer climates.

Symptoms

Fruit rot symptoms differ somewhat, depending on whether infection is initiated by ascospores from the sexual stage or conidia of *C. acutatum* and *C. gloeosporioides*, which produce only conidia. Initial symptoms produced by either strain are similar.

Lesions begin as small, slightly sunken areas that are light brown to dark brown. On mature fruit, lesions may be surrounded by red halos (*Photo 36*). Lesions originating from infections by *C. acutatum* or *C. gloeosporioides* remain circular and become sunken as they enlarge.

Copious quantities of conidia are produced in acervuli, which occur in concentric circles around the point of infection (*Photos 36–39*). Acervuli are sparse on some lesions and very dense on

others. Under moist, humid conditions, the spore masses appear creamy and are salmon to pink.

Lesions initiated by ascospores of *G. cingulata* are usually not sunken and are often darker brown than those caused by conidia of *C. gloeosporioides* or *C. acutatum*. Acervuli are widely scattered over the surface, and perithecia are found in dark brown to black clumps scattered on the surface of lesions (*Photo 40*).

Lesions initiated by *Colletotrichum* spp. and *G. cingulata* extend in a cone shape toward the core. In cross-section, the

Photo 36. Golden Delicious apple with bitter rot lesion. Note the sunken nature and presence of acervuli with salmon or pink spores.

Photo 37. Later stage of bitter rot. Note depressed lesions with concentric rings of acervuli and conidia.

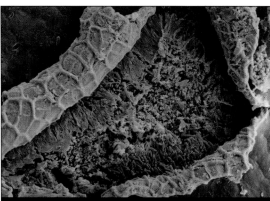

Photo 38. Conidia of *Colletotrichum gloeosporioides* within an acervulus.

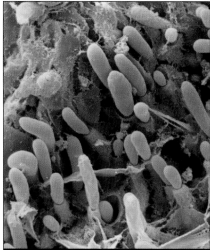

Photo 39. Conidia of *Colletotrichum gloeosporioides* extruding from phialides within the acervulus.

Photo 40. Golden Delicious apple infected with a perithecial strain of the bitter rot fungus. Lesions are usually not sunken and are often darker than those caused by conidial strains.

lesion appears V-shaped (*Photo 33*). This is a reasonably reliable characteristic that can be used to distinguish bitter rot from white rot or black rot. The rotten area is brown but much firmer than areas affected by white rot. Infected fruit mummify, and some may remain attached to the tree through the winter.

A leaf spot has been associated with *G. cingulata*. Spots begin as small, red flecks, which enlarge to irregular, brown spots 1⁄16 to 1⁄2 inch in diameter. Severely affected leaves may abscise.

Bitter rot cankers are rare in the eastern United States. Cankers are oval, sunken and often marked with zones or concentric rings, like a target.

Disease Cycle

Glomerella cingulata and *Colletotrichum* spp. primarily over-winter in dead wood or mummies in the tree. Other inoculum sources include stems of fruit that were torn from them at harvest or fruit mummified by chemical thinners. Conidia produced in these overwintering sites are the primary inoculum source in the spring, although ascospore inoculum is important in some orchards. Conidia are spread primarily by rain. Ascospores are released by rain and are airborne. Fruit are susceptible from 3 weeks after petal fall until harvest. Temperatures of 80 to 90 degrees F are most favorable for disease development. Because of the large number of conidia produced in lesions on fruit and the rapid disease cycle, spread of the disease within the orchard can be very rapid.

Control

Removing mummified fruit, dead wood and fire-blighted twigs is important to help control the disease. Fungicides, applied on a 10- to 14-day schedule from petal fall until harvest, are the most important means of control. More frequent applications may be necessary under conditions favorable for disease development. Removing newly infected fruit from trees during the growing season will aid in control. Although apple cultivars do not vary widely in their susceptibility, the disease is often more severe on Empire, Freedom, Golden Delicious, Fuji, Granny Smith and Arkansas Black.

Selected References

Bernstein, B., E. I. Zehr, R. A. Dean and E. Shabi. 1995. Characteristics of *Colletotrichum* from peach, apple, pecan and other hosts. *Plant Dis.*, 79:478–482.

Noe, J. P., and J. E. Starkey. 1982. Relationship of apple fruit maturity and inoculum concentration to infection by *Glomerella cingulata*. *Plant Dis.*, 66:379–381.

Shane, W. W., and T. B. Sutton. 1981. Germination, appressorium formation, and infection of immature and mature apple fruit by *Glomerella cingulata*. *Phytopathology*, 71:454–457.

Sutton, T. B., and W. W. Shane. 1983. Epidemiology of the perfect stage of *Glomerella cingulata* on apples. *Phytopathology*, 73:1179–1183.

Pome Fruits

SOOTY BLOTCH AND FLYSPECK

Sooty blotch and fly-speck are common names of two diseases that often occur on apple and pear fruit at the same time. Sooty blotch is a disease complex caused by *Peltaster fructicola* Johnson, *Geastrumia polystigmatis* Batista and M. L. Farr, *Leptodontium elatius* (G. Mangenot) De Hoog and other fungi. Flyspeck is caused by *Zygophiala jamaicensis* E. Mason. Their presence on the fruit's surface lowers quality and subsequent market value. Both diseases are economically important problems in the southern United States and of occasional economic importance, particularly on late-maturing cultivars and on scab-resistant cultivars grown without fungicides, in the northeast and north central regions.

Symptoms

The two diseases are recognized primarily by their macroscopic signs. When they occur together on fruit, the colonies are mutually exclusive.

Sooty blotch appears as superficial sooty or cloudy blotches on the surface of fruit. Blotches are brown to olive green and indefinite in outline (*Photo 41*). The fungi that cause sooty blotch produce clusters of colonies that range in appearance from sooty and smudge-like to much darker colonies with many small, circular pycnidia scattered within them. All of the fungi that cause sooty blotch are usually found within an orchard, but the predominant type may vary from orchard to orchard.

Flyspeck appears on fruit as sharply defined, black, shiny dots in groups of a few to nearly 100 (*Photo 42*). These dots or specks are sexual fruiting structures called pseudothecia. They are much larger than the pycnidia associated with the most common types of sooty blotch colonies.

Disease Cycle

Both pathogens overwinter on twigs of many woody plants. Spread of the sooty blotch fungus from overwintering hosts is by waterborne conidia or mycelial fragments. Primary infection by the flyspeck fungus is by airborne ascospores, which are discharged during rainy periods. Secondary infections occur by conidia, which are airborne or waterborne. The two-celled ascospores of *Z. jamaicensis* can be found in the pseudothecia in the late spring and early summer. Fruit infection can occur anytime after petal fall but is most prevalent in mid- to late summer.

Both diseases are favored by temperatures that typically occur in summer in the South, when combined with high humidity and abundant rainfall. In northern states, disease outbreaks are favored by extended periods of above normal summer temperatures combined with frequent rainfall.

Control

Control is achieved through sanitation and the use of fungicides. Removing reservoir hosts, particularly brambles, from the orchard and surrounding hedgerows helps to reduce the influx of inoculum, but in wet years this by itself may not be adequate for disease control.

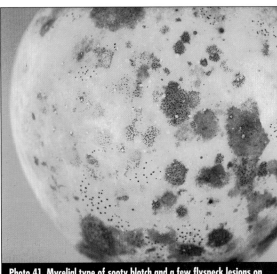

Photo 41. Mycelial type of sooty blotch and a few flyspeck lesions on the surface of apple.

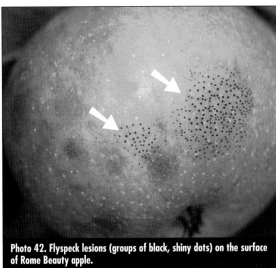

Photo 42. Flyspeck lesions (groups of black, shiny dots) on the surface of Rome Beauty apple.

Cultural practices that facilitate drying of fruit following rain or dew, such as dormant and summer pruning to open up the tree canopy and thinning to separate fruit clusters, help to prevent disease. These practices also favor better spray coverage and fruit quality. Both diseases are very difficult to control in orchards or in areas of orchards with restricted air movement.

Fungicide treatments are initiated 1 to 2 weeks after petal fall, then repeated on a 10- to 14-day schedule until 2 to 3 weeks before harvest. Alternatively, timing of the first application may be based on the number of hours of leaf wetting that have accumulated since 10 days after petal fall. In southern states, the first fungicide treatment is applied after the accumulation of 250 hours of leaf wetting of 3 hours' duration or longer beginning with the first rain that occurs within 10 days after petal fall. In northeastern states, the first spray for sooty blotch is probably not needed until 300 hours of leaf wetness have been accumulated.

Selected References

Baines, R. C., and M. W. Gardner. 1932. Pathogenicity and cultural characters of the apple sooty blotch fungus. *Phytopathology,* 22:937–952.

Baker, K. F., L. H. Davis, R. D. Durbin and W. C. Snyder. 1977. Greasy blotch of carnation and flyspeck of apple: Diseases caused by *Zygophiala jamaicensis. Phytopathology,* 67:580–588.

Brown, E. M., and T. B. Sutton. 1995. An empirical model for predicting first symptoms of sooty blotch and flyspeck of apples. *Plant Dis.,* 79:1165–1168.

Nasu, H., and H. Kunoh. 1993. The pathological anatomy of *Zygophiala jamaicensis* on fruit surfaces. Pages 137–155 in: *Cytology, Histology and Histochemistry of Fruit Tree Diseases.* A.R. Biggs, ed. Ann Arbor, Mich.: CRC Press.

Ocamb-Basu, C. M., T. B. Sutton and L.A. Nelson. 1988. The effects of pruning on incidence and severity of *Zygophiala jamaicensis* and *Gloeodes pomigena* infections of apple fruit. *Phytopathology,* 78:1004–1008.

Sutton, A. L., and T. B. Sutton. 1994. The distribution of the mycelial types of *Gloeodes pomigena* on apples in North Carolina and their relationship to environmental conditions. *Plant Dis.,* 78:668–673.

Sutton, T. B. 1990. Dispersal of conidia of *Zygophiala jamaicensis* in apple orchards. *Plant Dis.,* 74:643–646.

Sutton, T. B., J. J. Bond and C. M. Ocamb-Basu. 1988. Reservoir hosts of *Schizothyrium pomi,* cause of flyspeck of apples in North Carolina. *Plant Dis.,* 72:801.

Pome Fruits

ALTERNARIA BLOTCH

Alternaria blotch, caused by *Alternaria alternata* pathotype *mali* (*Alternaria mali*), has been a serious problem on apples in Japan and other countries in the Far East since the late 1950s. It was first reported in the United States in the late 1980s in North Carolina and has subsequently been reported from most states in the Southeast. The disease causes extensive losses in severely affected orchards of Delicious.

Symptoms

The disease is characterized by circular, brown necrotic spots ⅛ to ¼ inch in diameter on the leaves (*Photos 43, 44*). Leafspots often have purple borders. Symptoms are very similar to those associated with captan injury and frog-eye leaf spot (caused by *Botryosphaeria obtusa*) and are often confused with them. Leaves infected with *A. mali* are usually uniformly distributed on terminals throughout the tree and orchard; leaves with frog-eye leaf spots are localized and associated with dead wood or mummied apples, and those associated with captan injury often are localized on leaves with relatively high captan applications. Severely affected leaves abscise, and defoliation can approach 90 percent by harvest in severely affected orchards.

Fruit symptoms are relatively indistinct on the most susceptible cultivars grown in the United States and are limited to darkened and somewhat raised lenticels.

Disease Cycle

Alternaria alternata pathotype *mali* survives from one season to the next in fallen leaves and in buds, although leaves are the most important overwintering site. In mid-May to early June, conidia produced on the leaves are blown onto apple foliage where they initiate primary infection. Infection can occur in as few as 5 hours under favorable temperatures (70 to 80 degrees F) and moisture. Infections occur on leaves of all ages, but young leaves tend to become infected to a greater extent. Strains of Delicious and Empire are most seriously affected. Defoliation from the disease is more severe if high populations of mites are present.

Control

No chemicals are registered for the control of Alternaria blotch in the United States. Chopping leaves with a mower or removing them will help reduce the inoculum level the next spring. Mites should be maintained at or below established IPM thresholds.

Photo 43. Alternaria blotch on Delicious apple leaf.

Photo 44. Progressive development of Alternaria blotch across leaves on a shoot of Delicious apple.

Selected References

Filajdic, N., and T. B. Sutton. 1991. Identification and distribution of *Alternaria mali* on apples in North Carolina and susceptibility of different varieties of apples to Alternaria blotch. *Plant Dis.*, 75:1045–1048.

Filajdic, N., and T. B. Sutton. 1992. Influence of temperature and wetting duration on infection of apple leaves and virulence of different isolates of *Alternaria mali*. *Phytopathology*, 82:1279–1283.

Filajdic, N., and T. B. Sutton. 1995. Overwintering of *Alternaria mali*, the causal agent of Alternaria blotch of apple. *Plant Dis.*, 79:695–698.

Filajdic, N., T. B. Sutton, J. F. Walgenback and C. R. Unrath. 1995. The influence of European red mites on intensity of Alternaria blotch of apple and fruit quality and yield. *Plant Dis.*, 79:683–690.

PEAR SCAB

Pear scab, caused by *Venturia pirina* Aderhold, occurs sporadically throughout pear-growing areas of the eastern United States. Once established in an orchard, it can cause serious economic loss by reducing the appearance and quality of the fruit and sometimes crop size.

Photo 45. Scab on young pear fruit and leaves.

Photo 46. Pear scab lesions on current-season shoot growth.

Symptoms

Pear scab occurs on the leaves, shoots and fruit (*Photos 45, 46*). The symptoms are similar to those of apple scab except that shoot infections are common on pear but rare on apple.

Fruit infections are initially olivaceous, velvety, circular spots; they become black and corky later in the season. The skin on infected fruit often cracks, causing distortion of the fruit as they mature. The fruit may drop during the summer if lesions are present on fruit pedicels. Fruit infections initiated late in the summer form rough, black, circular lesions some weeks after the fruit are placed in storage.

Lesions on foliage are most evident on the lower surfaces of leaves. These undersurface lesions are rather inconspicuous, olivaceous, circular spots. Conidia are produced in the new lesions.

Lesions on shoots resemble those found on fruit. Velvety, elliptical spots with abundant conidia are produced on the current season's growth. The lesions become dark brown to black with age and eventually slough off.

Disease Cycle

Ascospores from pseudothecia in leaves on the orchard floor and conidia from twig lesions are the sources of primary inoculum in spring. The presence of ascospores is necessary for severe outbreaks of pear scab in the eastern United States.

Mature ascospores usually are present at bud swell and are released over a period of about 3 months. Ascospore maturity is a function of degree-days and, provided moisture is adequate, an accumulation of about 1,800 degree-days (DD) at a base of 32 degrees F (1,000 DD base 0 degrees C) is required to mature all the spores. Rain or heavy dew is required for ascospore release. Although ascospores may be released with less than 0.01 inch of rain or with dew, the duration of wetness associated with these events is usually not sufficient for infection. Ascospore release is favored by light but can also occur in darkness; however, the magnitude of daytime release is greater than that at night. Maximum release occurs between early bloom and petal fall.

Infection of pear foliage and fruit is governed by the number of conidia and ascospores in the orchard and by the temperature and wetness duration during periods when the inoculum is being disseminated. Incidence and severity of scab generally increase directly with increases in inoculum concentration. Minimum wetness duration for foliar infection varies from 10 hours at 75 degrees F to 25 hours at 45 degrees F and is similar to that required for light infection with *V. inaequalis* ascospores as defined by Mills. The incubation period for pear scab is 10 to 19 days, slightly longer than that for apple scab. Pear fruit appear to be as susceptible as leaves to infection when young, but they become increasingly resistant to infection as they mature.

Strains of the pear scab fungus differ in the cultivars they infect. Resistant cultivars in one region may not be resistant in another, depending on the types and abundance of pear scab strains present.

Control

Many of the fungicides that control apple scab also control pear scab, but the spray program for pear scab control is often minimal because pear scab is seldom as severe as apple scab. Protective sprays are initiated when green tissue emerges from the buds, and they are repeated on a 7- to 10-day interval until the supply of ascospores is exhausted. Postinfection control of pear scab may be possible with the use of certain fungicides with "kick-back" action within a critical time following infection. Duration of wetness periods and temperature data, along with the Mills table (*Table 1*), are used to determine when infection occurs. A second postinfection application is applied 7 days after the first one.

Selected References

Latorre, B. A., P. Yanez and E. Rauld. 1985. Factors affecting release of ascospores by the pear scab fungus (*Venturia pirina*). *Plant Dis.,* 69:213–216.

Shabi, E., J. Rotem and G. Loebenstein. 1973. Physiological races of *Venturia pirina* on pear. *Phytopathology,* 63:41–43.

Spotts, R. A., and L. A. Cervantes. 1991. Effect of temperature and wetness on infection of pear by *Venturia pirina* and the relationship between preharvest inoculation and storage scab. *Plant Dis.,* 75:1204–1207.

Spotts, R. A., and L. A. Cervantes. 1994. Factors affecting maturation and release of ascospores of *Venturia pirina* in Oregon. *Phytopathology,* 84:260–264.

Pome Fruits

FABRAEA LEAF SPOT

Fabraea leaf spot, caused by *Diplocarpon mespili* (Sorauer) Sutton (formerly *Fabraea maculata* Atk.), occurs on pear and quince. The disease occurs less commonly than pear scab but is more common than Mycosphaerella leaf spot. It has been known to build up rapidly in orchards where it had not been noticed for years.

Symptoms

Leaves, shoots and fruit can be infected. Where the disease is severe, substantial defoliation of pear trees may occur late in the summer and the fruit may be cracked, misshapen and unmarketable.

Infections first appear as small, purplish black dots and gradually enlarge to form circular, brown lesions about ⅛ to ¼ inch in diameter (*Photo 47*). A small, black pimple, or acervulus, develops in the center of each lesion, from which conidia ooze in a creamy white mass in wet weather (*Photo 48*). The conidia, as they ooze from acervuli, are distinctive. They are composed of four cells–two small lateral cells are present on either side of the juncture of two larger ones. Setae (appendages) attached to the apex of each cell give the spores an insectlike appearance (*Photo 49*). Lesions on new shoots appear as purple spots with indefinite margins. A few of these lesions may form superficial cankers that persist into the next growing season before being walled off.

Disease Cycle

Ascospores formed in apothecia in fallen leaves on the orchard floor and conidia formed in acervuli in cankers on shoots are the sources of primary inoculum. In the northeastern states, ascospore maturation can vary from mid–May to late July, depending on the year. Rainy periods promote ascospore discharge and new infections. In the southeastern states, where overwintering on shoots is important, conidia in cankers mature in late April or May and are disseminated by driving rain.

The minimum length of wetting required for infection is 12 hours at 50 degrees F or 8 hours at 68 to 77 degrees F. Lesions begin to appear about 7 days after the beginning of an infection period. Secondary conidia are produced in the primary lesions, and the disease may advance rapidly in late summer as rain and wind distribute the conidia throughout the tree.

Photo 47. Fruit of Tennessee Selection #345298 (left) and Bartlett pear (right) infected by the Fabraea leaf spot pathogen. (Courtesy [right] D. A. Rosenberger, Hudson Valley Laboratory, Cornell University, Highland, New York)

Photo 48. Surface view of an acervulus of *Entomosporium maculatum* breaking through the epidermis.

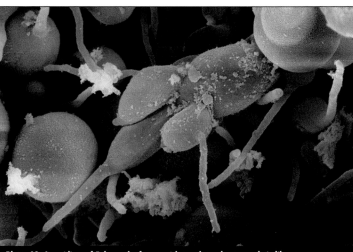

Photo 49. A conidium of Fabraea leaf spot with two lateral setae or hair-like structures attached to the cells.

Control

Fabraea leaf spot is controlled by applying protectant fungicides. The timing and number of applications required depend on the source and availability of primary inoculum, which varies between regions in the eastern United States. Early–season spray programs for pear scab should also give initial control of Fabraea leaf spot. Where ascospores and conidia of *Fabraea* occur after petal fall, summer fungicide treatments are needed. In the northeastern states, fungicide applications in June and July will generally control Fabraea leaf spot, but mid–August and September applications are advisable in wet seasons on late cultivars such as Bosc.

Selected References

Bell, R. L., and T. van der Zwet. 1988. Susceptibility of *Pyrus* germplasm to Fabraea leaf spot. *Acta Hortic.*, 224:229–236.

Goldsworth, M. C., and M. A. Smith. 1938. The comparative importance of leaves and twigs as overwintering infection sources of the pear leaf blight pathogen, *Fabraea maculata. Phytopathology,* 28:572–582.

van der Zwet, T., and H. F. Stroo. 1985. Effects of cultural conditions on sporulation, germination and pathogenicity of *Entomosporium maculatum. Phytopathology,* 75:94–97.

MYCOSPHAERELLA LEAF SPOT

Mycosphaerella leaf spot, caused by *Mycosphaerella pyri* (Auersw.) Boerema, is a minor disease of pear. It rarely causes economic loss in fungicide–sprayed orchards, but the fungus may build up and cause early defoliation in unsprayed orchards.

Symptoms and Disease Cycle

Leaves and fruit can be infected. Leaf spots are grayish with purplish margins and contain several small, black pycnidia in their centers. When lesions are numerous, the leaves may become chlorotic and defoliation occurs. Lesions on fruit appear as small brown to black blisters *(Photo 50)*.

Ascospores from pseudothecia in leaves on the orchard floor are the source of primary inoculum. They are discharged during rain in spring. Conidia from pycnidia in leaf spots are disseminated during the summer by rain. Infection of leaves occurs through stomata; infection of fruit may occur directly through the epidermis. The disease is favored in years with above normal rainfall in summer.

Control

The control program for Fabraea leaf spot usually controls Mycosphaerella leaf spot as well.

Photo 50. Mycosphaerella leaf spot on pear leaves and fruit.

Selected References

Tzavella-Klonari, K., and D. Tamoutseli. 1986. The development and structure of the spermogonia and ascocarps of *Mycosphaerella sentina* (Fr.) Schroet. *Cryptogam. Mycol.,* 7:267–273.

Sivanesan, A. 1990. *Mycosphaerella pyri.* CMI Descriptions of Pathogenic Fungi and Bacteria, No. 989. Wallingford, U.K.: CAB International.

Pome Fruits

BLISTER SPOT

Blister spot of apple is caused by the bacterium *Pseudomonas syringae* pv. *papulans* (Rose) Dhanvantari. The disease occurs on the cultivar Mutsu (Crispin) in eastern North America from Georgia to Ontario, Canada, and west to Ohio and Michigan. Although the disease has been reported on numerous other cultivars, it is seldom of economic importance. Recently, minor outbreaks of blister spot were identified on the cultivars RedCort and Fuji.

Symptoms

Raised, brown to black blisters (*Photo 51*) appear on the fruit in mid- to late June in southern fruit-growing areas and mid- to late July in northern areas. Young lesions may appear water-soaked, and masses of rod-shaped motile bacteria can be observed microscopically in sections of the blistered tissue. Older lesions appear to erupt as the fruit expand and mature (*Photo 52*). Lesions are associated with lenticels, are shallow and ⅛ to ¼ inch in diameter. A few to more than 100 lesions may develop on each fruit. A red to purple zone of discoloration often develops in the skin bordering older lesions.

Affected leaves are curled, puckered and misshapen. A midvein necrosis with water soaking, infrequently with ooze droplets, may be observed on young leaves of Mutsu at or slightly before the onset of fruit symptoms (*Photo 53*). On older leaves, the midvein lesions are crusty brown.

Disease Cycle

The bacteria overwinter in apple buds and leaf scars. In spring and summer, rain can disseminate the bacteria to leaves, blossoms and fruit surfaces, where the bacteria multiply profusely without causing symptoms. Starting 2 to 2½ weeks after petal fall and continuing for 2 to 4 weeks, the fruit are highly susceptible to infection through natural openings called stomata. Lesions should appear on the fruit about 10 to 12 days after initial infection. The stomata are lost as the fruit expand, reducing the potential for infection.

Natural infection of Mutsu may involve a few to nearly all of the fruit in an orchard. Although natural infection is rarely observed on other cultivars, the disease has been produced by artificial inoculation on such common cultivars as Jonathan, Delicious, Golden Delicious and Rome Beauty.

Control

Problems with blister spot can usually be prevented when the orchard is established by avoiding the cultivar Mutsu. Until the development of streptomycin-resistant strains of the pathogen, the disease was controlled with well timed sprays of streptomycin during the period when the fruit are susceptible. In orchards without resistant strains, the first spray should be applied no later than 2 weeks after petal fall and should be followed weekly by two additional sprays. Resistant strains develop after only a few years of streptomycin usage. Once resistance develops, this control strategy is ineffective and further use of streptomycin only increases the resistance problem.

Photo 51. Shiny, raised blister spot lesions on Mutsu apple. (Courtesy W. G. Bonn, Agriculture Canada, Harrow, Ontario)

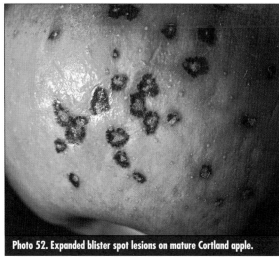

Photo 52. Expanded blister spot lesions on mature Cortland apple.

Photo 53. Shoot of Mutsu with ooze droplet containing blister spot bacteria and curled leaves due to infection on the midvein. (Courtesy W. G. Bonn, Agriculture Canada, Harrow, Ontario)

Selected References

Bonn, W. G., and K. E. Bedford. 1986. Midvein necrosis of Mutsu apple caused by *Pseudomonas syringae* pv. *papulans*. *Can. J. Plant Pathol.*, 8:167–169.

Burr, T. J., and B. Hurwitz. 1981. Seasonal susceptibility of Mutsu apples to *Pseudomonas syringae* pv. *papulans*. *Plant Dis.*, 65:334–336.

Burr, T. J., and B. Katz. 1984. Overwintering and distribution pattern of *Pseudomonas syringae* pv. *papulans* and pv. *syringae* in apple buds. *Plant Dis.*, 68:383–385.

Burr, T. H., J. L. Norelli, B. Katz, W. F. Wilcox and S. A. Hoying. 1988. Streptomycin resistance of *Pseudomonas syringae* pv. *papulans* in apple orchards and its association with a conjugative plasmid. *Phytopathology*, 78:410–413.

Jones, A. L., J. L. Norelli and G. R. Ehret. 1991. Detection of streptomycin-resistant *Pseudomonas syringae* pv. *papulans* in Michigan apple orchards. *Plant Dis.*, 75:529–531.

Pome Fruits

CALYX-END AND DRY-EYE ROTS

Calyx-end rot is caused by *Sclerotinia sclerotiorum* (Lib.) de Bary, and dry-eye rot by *Botrytis cinerea* Pers. The diseases occur separately, but their symptoms are virtually identical. Both diseases occur sporadically across the northeastern apple-growing region from Maine to Wisconsin. They seldom cause serious economic loss.

Symptoms and Disease Cycle

Affected fruit become visible in the orchard about 1 month after petal fall. A small, ¼- to ½-inch area of brown, rotted tissue develops adjacent to the blossom end of the fruit (*Photo 54*). If the disease is caused by *S. sclerotiorum*, the rot is soft and often expands until one-third or more of the fruit is involved (*Photo 55*). The rot associated with *Botrytis* infection is dry and shallow, and the rotted area is often surrounded by a red border.

Presumably, *S. sclerotiorum* overwinters as sclerotia associated with dandelion, wild clover and possibly other plants in the orchard sod. Apothecia are produced in early spring and release ascospores from early bloom to about 3 weeks after petal fall. During wet periods with suitable temperatures, the ascospores infect the blossoms and young fruit. These same environmental conditions also favor infection by conidia and ascospores of *B. cinerea*.

These diseases are noted most often on Paulared, McIntosh, Rome Beauty and Delicious, although other cultivars are undoubtedly infected.

Control

Calyx-end and dry-eye rots have not caused sufficient damage to justify a study of control measures. Some fungicides used to control apple scab may help prevent infection.

Prolonged storage of apples from orchards with a high incidence of infection may be undesirable because of a potential increase in secondary rot problems.

Selected References

Abawi, G. S., and R. G. Grogan. 1975. Source of primary inoculum and effects of temperature and moisture on infection of beans by *Whetzelinia sclerotiorum*. *Phytopathology*, 65:300–309.

Palmiter, D. H. 1951. A blossom-end rot of apples in New York caused by *Botrytis*. *Plant Dis. Rep.*, 35:435–436.

Rich, A. E. 1970. Calyx-end rot of apples. *Phytopathology*, 60:1152.

Tronsmo, A., and J. Raa. 1977. Life cycle of the dry-eye rot pathogen, *Botrytis cinerea*, on apple. *Phytopathology Z.*, 89:203–207.

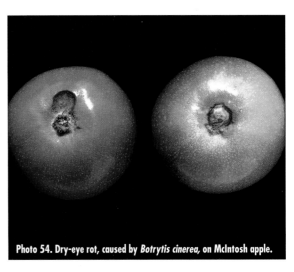

Photo 54. Dry-eye rot, caused by *Botrytis cinerea*, on McIntosh apple.

Photo 55. Calyx-end rot on Paulared apple.

BLACK POX

Black pox, caused by the fungus *Helmintho-sporium papulosum* A. Berg, affects apple twigs, fruit and foliage. It is most severe in the southern Appalachian growing area but has been reported from most mid–Atlantic and lower midwestern states. The disease also affects pear.

Symptoms

On bark, infections begin as well defined conical, shiny black elevations called papules. Secondary growth of the fungus results in the formation of necrotic, dark, sunken areas ¼ to ½ inch in diameter. Because bark becomes resistant with age, infections on older twigs are small and may not show any secondary enlargements. Symptoms of twigs infected with *H. papulosum* may be confused with internal bark necrosis, a physiological disorder. Black pox may be identified by incubating infected twigs in a moist chamber and examining the margins of the papules or bark cankers for conidia.

Fruit infections appear as smooth, black, slightly sunken, round spots ⅛ to ⅜ inch in diameter (*Photo 56*).

Leaf spots begin as small (¹⁄₁₆ inch), light green lesions surrounded with red halos. Lesions enlarge to ¹⁄₁₆ to ¼ inch in diameter, and the centers become tan to brown.

Photo 56. Small black spots of black pox, caused by *Helminthosporium papulosum*, on Grimes Golden apples.

Disease Cycle

Helminthosporium papulosum overwinters in twig cankers. Conidia produced on these cankers are blown or washed onto leaves, twigs and fruit. Bark on twigs remains susceptible for several years. Rainy periods with temperatures of 70 to 80 degrees F are most favorable for infection.

Control

Disease-free planting stock is important to help prevent the spread of the disease. Most fungicides used in the cover sprays at 14–day intervals for summer diseases satisfactorily control black pox. The disease is often more severe on Golden Delicious than other cultivars.

Selected References

Berg, A. 1933. *Black pox and other apple bark diseases commonly known as measles.* Bull. 260. Blacksburg, Va.: West Virginia Agricultural Experiment Station.

Taylor, J. 1963. An apple leaf spot caused by *Helminthosporium papulosum. Plant Dis. Rep.,* 47:1105–1106.

Pome Fruits

BROOKS SPOT

Brooks spot, caused by the fungus *Mycosphaerella pomi* (Pass.) Lindau, occurs in most of the apple-growing areas in the eastern United States. It is most severe in the southeastern mid–Atlantic and northeastern apple–growing regions. Although it's considered a minor disease, it has affected more than 50 percent of the fruit in some orchards. The disease is also known as fruit spot or Phoma fruit spot. When the disease occurs on quince, it is called quince blotch.

Symptoms

Brooks spot appears on immature fruit in late June or July as irregular, slightly sunken, dark green lesions. As fruit ripen on red–skinned cultivars, the lesions become dark red to purple to almost black. On light–skinned cultivars, lesions remain dark green (*Photos 57, 58*). Lesions are generally more common on the calyx end. Fruit may crack if infections occur early and are numerous. Lesions may be confused with those of cork spot or bitter pit, but unlike those of cork spot or bitter pit, lesions of Brooks spot are shallow and the flesh of the fruit is not corky beneath the lesions. The fungus can be easily isolated from fruit lesions on water agar and forms the *Cylindrosporium* stage. Lesions on leaves appear as small purple spots (*Photo 59*).

Disease Cycle

Mycosphaerella pomi overwinters in apple leaves on the orchard floor. Pseudothecia form during the winter and spring, and ascospores are matured and discharged during rainy periods in May and June. Conditions favoring infection have not been precisely defined. Jonathan, Stayman, Rome Beauty and Golden Delicious are susceptible. Delicious is somewhat resistant.

Control

Most of the fungicides used in the early cover sprays for summer diseases adequately control Brooks spot. The sterol demethylation inhibitor (DMI) fungicides used for scab control are not effective.

Selected References

Brooks, C., and C. A. Black. 1912. Apple fruit spot and quince blotch. *Phytopathology*, 2:63–72.

Sutton, T. B., E. M. Brown and D. J. Hawthorne. 1987. Biology and epidemiology of *Mycosphaerella pomi*, cause of Brooks fruit spot of apple. *Phytopathology*, 77:431–437.

Yoder, K. S. 1982. Fungicide control of Brooks fruit spot of apple. *Plant Dis.*, 66:564–566.

Photo 57. Brooks spot on Jonathan apple fruit. (Courtesy K. S. Yoder, Agricultural Research and Extension Center, Virginia Polytechnic Institute, Winchester)

Photo 58. Brooks spot at the calyx end of Grimes Golden apples.

Photo 59. Small purple spots on an apple leaf caused by the Brooks spot fungus.

Pome Fruits

BLOTCH

Blotch, caused by the fungus *Phyllosticta solitaria* Ell. & Ev., was at one time considered second only to scab as the most important apple disease in the eastern United States from Pennsylvania, Indiana and Ohio southward. Today, the disease rarely occurs in commercial apple plantings.

Symptoms

Blotch affects leaves, buds, twigs and fruit. Interveinal lesions on leaves begin as circular, yellowish green spots about 1/16 inch in diameter. A single pycnidium with unicellular, ovoid and hyaline (transparent) conidia often forms in the center of each spot. As the spots age, they turn light tan to white. Spots on leaf veins and petioles are often much larger and may girdle the petioles and cause defoliation. Several hundred spots may occur on a single leaf.

Twig and branch cankers are often located at leaf nodes or at the bases of spurs that have developed from dormant buds. Infections on the current season's growth appear in late summer. Young cankers are dark, purplish or black, and raised or blister-like. During the second season, cankers enlarge and become light tan or orange. Older cankers become roughened as they enlarge, and bark is sloughed off. Cankers may coalesce, girdling twigs and small branches. Pycnidia are produced in cankers.

Fruit infections begin as small, dark, somewhat raised blisterlike areas. Lesions enlarge slowly and may reach 1/2 inch or more in

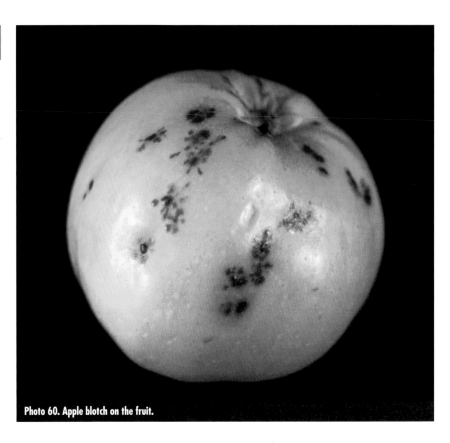
Photo 60. Apple blotch on the fruit.

diameter. Lesions are slightly raised with irregular or lobed margins (*Photo 60*). As lesions age, they turn from dark brown to black. Infected tissues harden and may crack. Pycnidia may be produced in the lesions.

Disease Cycle

Phyllosticta solitaria overwinters in twig and branch cankers or infected fruit. Conidia produced in pycnidia in cankers are washed or blown onto fruit, foliage and twigs. Fruit infection can occur anytime from petal fall until harvest, although fruit become more resistant as they mature. New leaf and twig growth, particularly water sprouts, are susceptible throughout the growing period. The optimum temperature for infection is 77 to 86 degrees F. Old cultivars such as Duchess, Ben Davis, Black Twig and Yellow Transparent are very susceptible. Delicious and Winesap are resistant. The planting of resistant cultivars is one of the main reasons this disease is no longer important.

Control

Control is achieved though planting disease-free nursery stock, planting resistant cultivars such as Delicious and following a regular spray program.

Selected References

Guba, E. F. 1924. Phyllosticta leaf spot, fruit spot and canker of apple—its etiology and control. *Phytopathology*, 14:234–237.

Kohl, E. J. 1932. Investigation on apple blotch. *Phytopathology*, 22:349–369.

Pome Fruits

X-SPOT

X–spot occurs in the southern Appalachian apple–growing region. The causal organism has not been identified, although *Nigrospora oryzae* (Berk. and Broome) Petch has been associated with the disease.

Symptoms and Disease Cycle

X–spot is characterized by small, circular, shallow, light tan, necrotic spots that often occur at the calyx end of the fruit (*Photo 61*). The disease is most common late in the season.

Factors affecting disease development are not known.

Control

Fungicides applied for summer diseases give effective control of X–spot.

Selected References

Yoder, K. S. 1990. X-spot. Page 28 in: *Compendium of Apple and Pear Disease.* A. L. Jones and H. S. Aldwinckle, eds. St. Paul, Minn.: American Phytopathological Society.

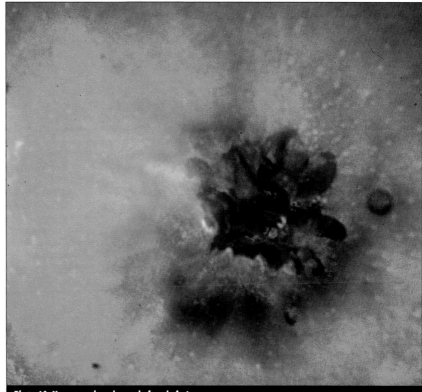

Photo 61. X-spot on the calyx end of apple fruit.
(Courtesy K. S. Yoder, Agricultural Research and Extension Center, Virginia Polytechnic Institute, Winchester)

Pome Fruits

APPLE MOSAIC

This virus disease was once very common in apple orchards because it was transmitted through nursery stock. Nurserymen now use budwood free of the apple mosaic virus. Symptoms occur as creamy yellow areas in leaves on one to a few branches of the tree (*Photo 62*). This disease is found mainly in old orchards and is included in this publication because the symptoms are sometimes confused with damage from herbicides.

Selected References

Gilmer, R. M. 1958. Two viruses that induce mosaic of apple. *Phytopathology,* 48:432–434.

Photo 62. Creamy-yellow areas in apple leaves infected with apple mosaic virus.

Pome Fruits

STONY PIT OF PEAR

Stony pit, a virus disease, severely misshapes and gnarls pear fruit, making them unsaleable *(Photo 63)*. Less severely infected fruit may be dimpled, a symptom often confused with plant bug injury. Grit cells concentrate beneath pits and make it difficult to cut the fruit with a knife. A rather pronounced, roughened bark or measles-like symptom has been associated with some strains of the stony pit virus. Leaf symptoms are known but are difficult to distinguish in the orchard. Bosc is most severely affected; Anjou, Winter Neils and Waite are affected to a lesser extent.

Control

Select trees known to be virus-free to prevent problems with this virus in new plantings. Though it is best to remove infected trees, removing all infected Bosc trees in some orchards could reduce Bartlett pear yields because of lack of sufficient pollenizers.

Photo 63. Bosc pear fruit severely gnarled from stony pit.

Selected References

Jones, A. L. 1964. Prevalence of stony pit virus in Bosc pears in western New York State. *Plant Dis. Rep.,* 48:385–387.

Kienholz, J. R. 1939. Stony pit, a transmissible disease of pears. *Phytopathology,* 29:260–267.

Thomson, A. 1989. Stony pit. Pages 182–187 in: *Virus and Virus-like Diseases of Pome Fruits and Simulating Noninfectious Disorders.* P. R. Fridlund, ed. Spec. Publ. SP003. Pullman, Wash.: Washington State University Cooperative Extension Service.

Photo 64. Dieback of Rome Beauty apple shoot from Nectria twig blight. Note canker at the base of the shoot.

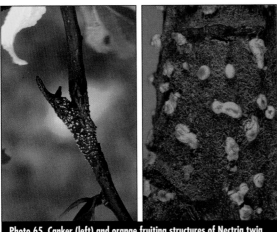

Photo 65. Canker (left) and orange fruiting structures of Nectria twig blight fungus (enlarged at right). (Courtesy [left] D. A. Rosenberger, Hudson Valley Laboratory, Cornell University, Highland, New York)

Photo 66. Orange fruiting structures of the Nectria twig blight fungus in a canker on McIntosh apple associated with a pruning wound.

NECTRIA TWIG BLIGHT

Nectria twig blight is a minor disease that results in dieback of apple twigs. It is caused by the fungus *Nectria cinnabarina* (Tode) Fr., asexual state of *Tubercularia vulgaris* Tode. Recognition of the disease is important because it is often confused with fire blight, which requires different control measures.

Symptoms and Disease Cycle

In June, shoot growth on infected twigs begins to wilt and die (*Photo 64*). Small, sunken cankers are found at the bases of the wilted shoots. Leaves on infected shoots appear to die from the base, not from the tip, as with fire blight, and no signs of blighted blossom clusters remain on the twigs. In mid- to late summer, bright orange or coral-red structures (sporodochia) ⅛ to ⅙ inch in diameter appear on the surface of the cankers (*Photo 65*). Orange sporodochia are also often associated with pruning wounds and winter-injured tissues on apple (*Photo 66*). In autumn, the fungus produces clusters of small, dark red, globular perithecia, but these structures are rarely present in the eastern United States.

Optimum fungal growth occurs in culture at 70 degrees F, with moderate growth at 80 to 85 degrees F and limited growth at 37 degrees F. Wounds from fruit harvest, which probably take a long time to heal late in the season, combined with prolonged periods of wet weather after harvest, appear to favor the establishment of infection. The disease has been noted primarily on cultivars with a large cluster-bud base, such as Rome Beauty, Ben Davis and Northern Spy.

Control

The disease is not usually severe enough to require special control measures. As a result, chemical control procedures have not been developed. Removal of infected twigs helps reduce the carryover of inoculum.

Selected References

Jones, A. L. 1963. Occurrence of Nectria twig blight of apple in western New York in 1962. *Plant Dis. Rep.*, 47:538–540.

Rosenberger, D. A., T. J. Burr and J. D. Gilpatrick. 1983. Failure of canker removal and postharvest fungicide sprays to control Nectria twig blight on apple. *Plant Dis.*, 67:15–17.

Thomas, H. E., and A. B. Burrell. 1929. A twig canker of apple caused by *Nectria cinnabarina*. *Phytopathology*, 19:1125–1128.

Pome Fruits

NECTRIA CANKER

Nectria canker, caused by *Nectria galligena* Bres., is occasionally found on apple nursery stock shipped into the eastern United States. The economic effects of the disease in the East are minor.

The fungus grows deep into the wood and kills new wound callus as it develops. This annual killing of successive layers of callus results in perennial, target–like, zonate cankers *(Photo 67)*. Eventually the cankers girdle the tree, resulting in a dieback of infected limbs or trees. The cankers are sometimes visible on the trunks of nursery trees at the time of planting, although latent infections can appear later in the season.

Nursery trees purchased from the western United States and Europe should be examined carefully for symptoms of the disease. Trees with cankers should be returned to the nursery for replacement or discarded.

Selected References

Dubin, H. J., and H. English. 1974. Factors affecting apple leaf scar infection by *Nectria galligena* conidia. *Phytopathology,* 64:1201–1203.

Swinburne, T. B. 1975. European canker of apple *(Nectria galligena). Rev. Plant Pathol.,* 54:789–799.

Photo 67. Nectria canker on the trunk of Delicious apple.

LEUCOSTOMA CANKER

Leucostoma canker, caused by *Leucostoma cincta* (Fr.) Höhn., occurs in Michigan as a minor disease on the cultivar Delicious and a rare disease on the cultivars Empire, Rome Beauty and Spartan.

Symptoms and Disease Cycle

Large, oval to elliptical cankers with very rough, scaly bark form at pruning cuts near the bases of scaffold limbs and the central leader *(Photo 68)*. Cankers can also start around ragged stubs at the bases of broken branches. As cankers expand they girdle limbs, causing scaffold limbs and even the central leader to wilt and die back during midseason *(Photo 69)*. Several black perithecia surrounded by a central pycnidium may be found in chambers in the bark, with black ostioles or pores opening to the surface.

Ascospores in the perithecia mature in early spring at about budbreak. They are probably released during rainy periods and disseminated to fresh injuries such as pruning wounds, where they infect.

Control

Removing and destroying the cankers will usually provide adequate control of this minor disease.

Photo 68. A Leucostoma canker on Delicious apple with characteristic rough, scaly bark.

Photo 69. Dieback of a central leader of Delicious apple caused by Leucostoma canker.

Selected References

Proffer, T. J., and A. L. Jones. 1989. A new canker disease of apple caused by *Leucostoma cincta* and other fungi associated with cankers on apple in Michigan. *Plant Dis.,* 73:508–514.

Surve-Iyer, R. S., G. C. Adams, A. F. Iezzoni and A. L. Jones. 1995. Isozyme detection and variation in *Leucostoma* species from *Prunus* and *Malus. Mycologia,* 87:471–482.

Pome Fruits

BLOSSOM BLAST OF PEAR

Blossom blast is caused by the bacterium *Pseudomonas syringae* pv. *syringae* van Hall. Although the disease has been reported from several northeastern and north central states, it occurs very sporadically and is of minor economic importance.

Symptoms and Disease Cycle

During cold, rainy periods around the bloom period, blossoms of pear sometimes are infected. The disease causes a blighting of the blossoms and young leaves. Symptoms are often restricted to the lower part of the tree or to trees in low-lying portions of the orchard. The disease is referred to as blossom blast because blighting is sudden and kills very young buds and flowers.

Symptoms of bacterial blast differ from those of fire blight in that bacterial ooze, a common characteristic of fire blight, is not found on the infected tissues. Also, blast infections rarely extend to the shoot growth (*Photo 70*). Immature fruit, particularly on pear, will sometimes develop water-soaked areas that become firm, black lesions.

Control

Effective control measures are lacking. Streptomycin spray applications starting early in the bloom period have shown promise on pears.

Photo 70. Blossom blast of pear. Note death of spur but failure of shoot to become infected.

Selected References

Parker, K. G., and W. H. Burkholder. 1950. *Pseudomonas syringae* Van Hall on apple and pear in New York State. *Plant Dis. Rep.,* 34:100–101.

Panagopoulos, C. G., and J. E. Cross. 1964. Blossom blight and related symptoms caused by *Pseudomonas syringae* van Hall on pear trees. *Annu. Rep. E. Malling Res. Stn., Kent, 1963,* A47:119–122.

Sands, D. C., and J. L. McIntyre. 1977. Possible methods to control pear blast, caused by *Pseudomonas syringae. Plant Dis. Rep.,* 61:311–312.

Pome Fruits

THREAD BLIGHT

Thread blight is caused by the fungus *Corticium stevensii* Burt. The disease is primarily a problem on apple in poorly managed orchards in the southeastern United States.

Symptoms and Disease Cycle

Thread blight symptoms are readily noticed in early summer. Leaves wilt and turn brown, usually in the interior or shaded portions of the tree (*Photo 71*). Dead, curled leaves cling to blighted branches, frequently in mid-branch, with unaffected leaves still appearing on both sides of the diseased area. In well managed orchards, the disease is usually not seen until after harvest, when the fungicidal spray program has been discontinued.

Positive diagnosis in the field is made by observing signs of the fungus. A sparse, white mycelial fan can be observed in advance of dead areas on partially blighted leaves. This mycelium can frequently be traced as fine white threads back to the leaf petiole and twigs. Abscised leaves may be tied to twigs and leaves by this network of threads and mycelium. The fungus is present on twigs and branches as silvery-tan rhizomorphs and white to tan sclerotia, which become hard and dark brown with age. Rhizomorphs may be up to $\frac{1}{16}$ inch wide and sclerotia up to $\frac{3}{16}$ inch long and $\frac{1}{8}$ inch thick. These fungal structures are superficial and can be scraped from the bark. The bark and wood of blighted branches do not appear adversely affected by the disease. Sclerotia and rhizomorphs can also grow superficially on fruit.

Information on the disease cycle of thread blight is limited. Sclerotia retain viability over the winter under eastern Kentucky conditions. Leaf blight usually appears first in the late spring on branches having sclerotia from the previous season. During the season, the fungus may grow from blighted to adjacent healthy leaves. The disease is more severe when trees are growing in a moist, shaded environment.

Control

Thread blight, once established in an orchard, is difficult to control using fungicides. Preventive fungicide sprays applied to trees prior to infection may protect the orchard from thread blight. Avoiding "hollows" and other shaded and poorly ventilated areas when selecting an orchard site should help prevent the disease. Under light disease pressure, pruning out blighted twigs and branches may provide adequate control. Pruning to promote better penetration of sunlight and air may also help.

Selected References

Wolf, F. A., and W. J. Bach. 1927. The thread blight disease caused by *Corticium koleroga* (Cooke) Höhn., on citrus and pomaceous plants. *Phytopathology*, 17:689–710.

Photo 71. Apple branch affected with thread blight showing dead leaves still attached.

Pome Fruits

NECROTIC LEAF BLOTCH

The cause of necrotic leaf blotch (NLB) of apple is not known. It is apparently a physiological disorder whose occurrence is related to air temperature, light intensity and soil moisture. A hormonal imbalance may be involved because symptoms are enhanced by gibberellins and reduced by abscisic acid. The disorder occurs worldwide on Golden Delicious and its bud sports. Seedlings out of Golden Delicious vary in susceptibility. Prime Gold and Nugget are also affected.

Symptoms and Disease Development

The disorder is characterized by the development of necrotic blotches or irregular areas of dead tissue in mature leaves (*Photo 72*). Midshoot leaves are most often affected (*Photo 73*). Blotches are limited by the leaf veins and vary greatly in size. Affected leaves begin to turn yellow after about 4 days and abscise a few days later. Some green leaves with NLB are also lost by abscission.

Two distinctive characteristics of NLB are that the symptoms develop suddenly, almost overnight, and in waves, generally from June through August. The disorder tends to be more common and severe later in the summer, however, and usually appears when a cool, rainy period is followed by hot summer weather. Some orchards or trees within an orchard may show little or no defoliation, while other orchards or trees reach 50 percent defoliation or more.

Amelioration

Although it appears that NLB is not caused by a fungus, bacterium or certain air pollutants, the disorder is reduced where the dithiocarbamate fungicides ziram or thiram are used in the summer spray program. Foliar applications of zinc oxide also have been effective in reducing the severity of the disorder.

Selected References

Sutton, T. B., and C. N. Clayton. 1974. *Necrotic leaf blotch of Golden Delicious apples*. Tech. Bull. 224. Raleigh, N.C.: North Carolina Agricultural Experiment Station.

Photo 72. Golden Delicious leaves with necrotic leaf blotch. Early symptom development is shown in top row; greatly advanced stage in bottom row. (Courtesy M. A. Ellis, Ohio Agricultural Research and Development Center, Wooster)

Photo 73. Leaf yellowing and defoliation of Golden Delicious from necrotic leaf blotch. The bottom leaves exhibit symptoms first, followed by upper leaves.

SOFT ROT/BLUE MOLD

Soft rot or blue mold is an economically important and common postharvest disease of apples and pears. It is caused by *Penicillium expansum* Link, *P. aurantiogriseum* Dierckx and a few other species of *Penicillium*.

Photo 74. Golden Delicious apple with soft rot/blue mold. Note cushions of blue-green fungus growth on surface lesion.

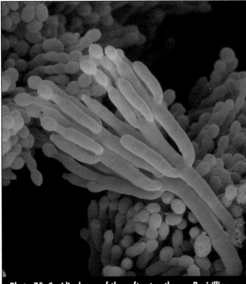

Photo 75. Conidiophores of the soft rot pathogen *Penicillium expansum*. Note one-celled ovoid conidia produced on brush-like structures.

Symptoms and Disease Cycle

Soft rot can usually be distinguished by its light color and the soft, watery texture of the decayed areas. The rot develops rapidly under favorable temperatures, often rotting the whole fruit in about 2 weeks. When humidity is high, the fungus develops gray-blue, cushion-like structures on the surface of the rot (*Photo 74*). These structures produce conidia important in spreading the disease (*Photo 75*).

The conidia are extremely resistant to drying and can survive on surfaces of packing and picking equipment. They are nearly always present to cause trouble when conditions favor infection and disease development. Conidia often build up in water used in dumping bulk boxes of fruit or in postharvest drenches for applying chemicals to inhibit storage scald.

Soft rot is a disease of ripe fruit, and it develops mostly on apples picked at an advanced state of maturity. It rarely occurs on immature fruit or in the orchard, except on fallen fruit. The pathogen infects the fruit through skin breaks and lenticels. Although stem punctures are usually considered the most important point of entrance, infection through lenticels in bruised areas may be serious at times.

Control

To reduce soft rot, harvest fruit at optimum maturity, handle carefully to prevent bruises and other injuries, and put into cold storage as soon as possible after harvest. If fruit are to be stored late into the season, use disinfectants or fungicides in the water during grading to reduce spread to healthy fruit. In many areas of the eastern United States, *Penicillium* has developed resistance to benzimidazole fungicides, and registrations for postharvest use on fruit crops have been withdrawn for most benzimidazole fungicides. Postharvest calcium treatment and biological control agents may replace these fungicides in the future.

Selected References

Conway, W. S., C. E. Sams, J. A. Abbott and B. D. Bruton. 1991. Postharvest calcium treatment of apple fruit to provide broad-spectrum protection against postharvest pathogens. *Plant Dis.*, 75:620–622.

Janisiewicz, W. J., D. L. Peterson and R. Bors. 1994. Control of storage decay of apples with *Sporobolomyces roseus*. *Plant Dis.*, 78:466–470.

Rosenberger, D. A., D. T. Wicklow, V. A. Korjagin and S. M Rondinaro. 1991. Pathogenicity and benzimidazole resistance in *Penicillium* species recovered from tanks in apple packing houses. *Plant Dis.*, 75:712–715.

Sanderson, P. G., and R. A. Spotts. 1995. Postharvest decay of winter pear and apple fruit caused by species of *Penicillium*. *Phytopathology*, 85:103–110.

Pome Fruits

GRAY MOLD

Gray mold of apple and pear, caused by *Botrytis* spp., is second in importance to soft rot as a decay or rot problem of fruit in cold storage. When fruit are removed from storage, gray mold–infected fruit will be almost completely decayed. Often, infection spreads from fruit to fruit during storage, producing "nests" or "pockets" of decayed fruit (*Photo 76*). Although gray mold infection may occur on individual fruit in the orchard and develop in storage, spores of the fungus often build up in solutions used to treat apples for scald or internal breakdown, and in water used in dumping bulk boxes. Adding an effective fungicide to these solutions will prevent spread of the disease.

Photo 76. A "nest" of apples with gray mold caused by *Botrytis*.
(Courtesy D. A. Rosenberger, Hudson Valley Laboratory, Cornell University, Highland, New York)

Selected References

Heald, F. D. 1926. A spot-rot of apples in storage caused by *Botrytis*. *Phytopathology*, 16:485–488.

DeKock, S. L., and G. Holz. 1991. Blossom-end rot of pears: Systemic infection of flowers and immature fruits by *Botrytis cinerea*. *J. Phytopathol.*, 135:317–327.

BULL'S-EYE ROT

Bull's–eye rot, or "northwestern anthracnose," is primarily a problem on fruit in storage, although infection originates in the orchard. It is a minor disease in the fruit–growing areas of the eastern United States. The disease is caused by the fungus *Pezicula malicorticis* (H. Jacks.) Nannf.

Symptoms

Small, elliptical, sunken cankers develop in the bark of twigs and branches that are less than 2 inches in diameter. Old cankers have a pronounced ridge of callus around the margin. New infections appear in the autumn as small, reddish brown areas on young twigs. As cankers enlarge during the autumn and spring, the bark remains smooth. Later, cracks develop between the canker and surrounding healthy tissue. Anthracnose cankers rarely enlarge after the first year.

Infected fruit develop brown, sunken, round lesions on the surface. The centers of the infected areas are often light–colored and surrounded by alternating areas of tan and brown, giving the characteristic "bull's–eye" appearance (*Photo 77*). Lesions tend to be relatively small. Whole fruit are rarely rotted by one lesion. The surface of the rot is sometimes covered with cream-colored spore masses.

Disease Cycle

Conidia of the bull's–eye rot fungus are washed by rain from cankers on twigs and branches to the apples and twigs below. Ascospores formed in 2–year–old cankers are forcibly discharged and disseminated by wind. Infection is mainly through the lenticels on the fruit, but wounds may also be infected. Fruit decay does not usually show up until after harvest.

The disease is generally more severe in years with frequent rains and moderate temperatures during the autumn and on fruit harvested beyond optimum maturity. Twig infections develop over a 2–year period. In the first autumn after infection, conidia are produced in acervuli on the surface of the canker; ascospores are produced in apothecia borne on cankers the second year.

Control

Removing twigs with cankers aids control by reducing the inoculum level in orchards. A fungicide spray applied in the dormant season and fungicide sprays near harvest help control fruit infection.

Selected References

McColloch, L. P., and A. J. Watson. 1966. Perennial canker of apples in West Virginia and Pennsylvania. *Plant Dis.*, 50:348–349.

Photo 77. Bull's-eye rot on McIntosh apple. Note light color in center of depressed lesion and alternating areas of tan and brown.

Pome Fruits

MOLDY CORE

Moldy core is primarily a disease of Delicious fruit. The problem originates in the orchard, where infected fruit sometimes color and ripen prematurely. More often, infected fruit appear normal when harvested and moldy core is not detected until prolonged storage. When infected fruit are cut in half, the seed cavity, or core region, is overgrown with mold (*Photos 78, 79*). A small proportion of the fruit may develop some rot in the flesh after prolonged storage or after the fruit are sold and held at room temperature.

Many genera of fungi can be isolated from the core region of apple. *Alternaria* spp. are the most common. Presumably, infection occurs through calyx tubes that remain open and allow spores to enter the seed cavity, where conditions are excellent for fungal development.

Control

There are no control measures. The incidence of moldy core has not been reduced by common fungicide spray programs, probably because once a fungus reaches the core region, it is protected against contact with fungicides.

Selected References

Carpenter, J. B. 1942. Moldy core of apples of Wisconsin. *Phytopathology*, 32:896–900.

Ellis, M. A., and J. G. Barrat. 1983. Colonization of Delicious apple fruits by *Alternaria* spp. and effect of fungicide sprays on moldy core. *Plant Dis.*, 67:150–152.

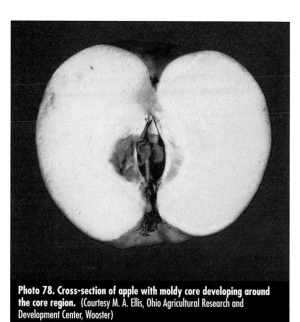

Photo 78. Cross-section of apple with moldy core developing around the core region. (Courtesy M. A. Ellis, Ohio Agricultural Research and Development Center, Wooster)

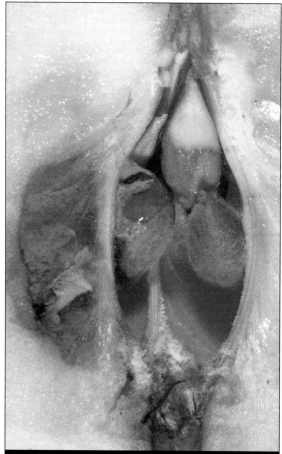

Photo 79. Close-up of seed cavity overgrown with mold. (Courtesy M. A. Ellis, Ohio Agricultural Research and Development Center, Wooster)

PHYTOPHTHORA ROOT, CROWN AND COLLAR ROT

Phytophthora root, crown and collar rot is caused by fungi in the genus *Phytophthora*. *Phytophthora cactorum* (Lebert and Cohn) Schröt. is often associated with the disease; however, *P. cambivora* (Petri) Buisman, *P. cryptogea* Pethybr. and Lafferty and a few other *Phytophthora* spp. have also been implicated in the disease. The importance of Phytophthora root, crown and collar rot has increased as size-controlling rootstocks have replaced seedling rootstocks.

Symptoms

The disease is characterized by cankers at or below the

Photo 80. Brown decay of rootstock collar just below ground, a common symptom of Phytophthora root, crown and collar rot.

ground line in the root–crown area (*Photo 80*). Infected bark becomes brown and is often slimy when wet. Cankers may extend from the point of origin into the root systems and up the trunk to the bud union. If the scion is susceptible to *Phytophthora*, the disease may extend above the union. Older cankers form a definite outline when the bark dries out, and callus tissues may develop at the margins.

Cankers girdle the roots and lower trunk, resulting in poor terminal growth and foliage discoloration. Severely infected trees eventually die. The leaves on affected trees may be small and yellow in summer or turn reddish to purple in early autumn. Because other root problems cause similar symptoms, these symptoms only indicate the need for further examination of the root system. Often, isolation of the fungus is required before a positive diagnosis can be made.

Young trees show the disease most when fruit production begins. This could be in the third year on some precocious rootstocks. Infected trees usually linger 2 or 3 years before dying. However, if conditions are optimum for infection, trees can be killed in one growing season.

Disease Cycle

The pathogen survives in soil for several years as oospores. These thick-walled structures can resist periods of unfavorable environment, such as drought, and are relatively resistant to chemical treatment. They occur in greatest numbers in old orchard soils. Infected or infested nursery stock is another important source of the fungus.

Fungal growth and infection are favored by damp, cool periods after harvest and in spring about the time leaves and flowers are emerging. The ability to produce many spores (primarily zoospores) allows the fungus to build up to high levels from a few oospores under favorable conditions. Zoospores swim in films of moisture to the crown or roots, where they initiate infection. Others may be splashed on the fruit and cause a rot. Rotten fruit are firm and light tan. Phytophthora fruit rot is most common in areas where the water used for overhead irrigation contains zoospores from infected forest and fruit trees.

Control

Phytophthora root, crown and collar rot can be reduced by carefully selecting the orchard site and rootstocks for the new planting. Susceptible rootstocks should not be planted in orchards (or areas of the orchard) with heavy, poorly drained soils. Where the soil is variable, it may be necessary to use several different rootstocks in the orchard, depending on their soil preference and susceptibility to collar rot. Tilling of wet areas in an otherwise well drained location often improves internal soil drainage sufficiently to avoid the disease in low areas. Planting on a raised soil bed greatly improves drainage around the tree and can reduce collar rot incidence. Mulching trees, if not done carefully, can increase the disease.

The clonal rootstocks vary in susceptibility to various species of *Phytophthora*. M.9 EMLA, Mark, Budagovsky (Bud.) 118 and Bud.9 are highly resistant; Bud.490, M.4, MM.111 EMLA, M.7 EMLA, M.26

EMLA and Polish (P.) 18 are inter-mediate; and MM.106 EMLA and Antonovka (Ant.) 313 are highly susceptible to *P. cactorum*. Mark and Bud.118 are highly resistant; Bud.9, M.7 EMLA and P.18 are intermediate, and the other rootstocks are susceptible to *P. cambivora*. M.4, MM.111 EMLA, Ant.313 and P.18 are susceptible to *P. cryptogea*. Except for seedlings of a highly susceptible cultivar such as Grimes Golden, seedling rootstocks have not exhibited significant susceptibility prob-lems to *Phytophthora* in the field. Since there are a number of *Phytophthora* spp. that cause root, crown and collar rots, it is dif-ficult to make absolute state-ments about the relative sus-ceptibility of different rootstocks to these diseases.

Several fungicides are registered for the control of Phytophthora root, crown and collar rot, but they are not a sub-stitute for good site preparation and the use of rootstocks adapted to the intended orchard site. Chemicals should be used on a preventive basis. Once infection occurs, it is difficult to eradicate the pathogen and save the tree.

Selected References

Browne, G. T., and S. M. Mircetich. 1993. Relative resistance of thirteen apple rootstocks to three species of *Phytophthora. Phytopathology,* 83:744–749.

Jeffers, S. N., and H. S. Aldwinckle. 1988. Phytophthora crown rot of apple trees: Sources of *Phytophthora cactorum* and *P. cambivora* as primary inoculum. *Phytopathology,* 78:328–335.

Jeffers, S. N., H. S. Aldwinckle, T. J. Burr and P. A. Arneson. 1982. *Phytophthora* and *Pythium* species associated with crown rot in New York apple orchards. *Phytopathology,* 72:533–538.

Merwin, I. A., W. F. Wilcox and W. C. Stiles. 1992. Influence of orchard ground cover management on the development of Phytophthora crown and root rots of apple. *Plant Dis.,* 76:199–205.

Wilcox, W. F. 1993. Incidence and severity of crown and root rots on four apple rootstocks following exposure to *Phytophthora* species and waterlogging. *J. Am. Soc. Hortic. Sci.,* 118:63–67.

Pome Fruits

WHITE ROOT ROT

White root rot is caused by the fungus *Scytinostroma galactinum* (Fr.) Donk. The disease occurs from Canada to Texas, although it is most prevalent in the southeastern apple–growing areas. Trees affected with white root rot usually die more rapidly than those affected with other root rots.

Symptoms and Disease Cycle

The surface of affected roots is covered with white or cream-colored mycelium. The mycelial growth is often thick and may extend into crevices in the soil close to roots (*Photo 81*). The cambium of affected roots is not always uniformly killed; as a result, zonate spots may be visible in the wood when the bark is removed. Affected roots are soft and lightweight.

Many forest trees and shrubs are hosts for *S. galactinum*. White oak is one of the most important hosts. Infection occurs when apple roots contact infected root pieces in the soil. The mycelium may remain viable on infected root pieces in the soil for many years.

Photo 81. White mycelium of the white root rot fungus growing over the surface of the trunk.

Control

No practical control is known. Stumps and roots of forest trees should be thoroughly removed from recently cleared sites. Replanting infested sites in established orchards is often not successful. If replanting is attempted, removal of all old roots is essential. No resistant rootstocks are available.

Selected References

Cooley, J. S., and R. W. Davidson. 1940. A white root rot of apple trees caused by *Corticium galactinum*. *Phytopathology,* 30:139–148.

Pome Fruits

BLACK ROOT ROT

Black root rot of apple is caused by the fungus *Xylaria mali* Fromme, particularly in the southern Appalachian states, and by *X. polymorpha* (Pers.) Grev. throughout the eastern United States. *Xylaria* has also been reported to cause a root rot of pear and sweet cherry.

Symptoms and Disease Cycle

The aboveground symptoms of trees affected with black root rot are similar to those caused by other root rot pathogens, vole injury, etc. Affected roots are brittle, easily broken and covered with a black fungal encrustation. Decayed wood is light and dry, and the bark remains firmly attached to the root. Trees with advanced root rot are easily uprooted. Fingerlike fruiting structures are occasionally found above ground at the bases of affected trees (*Photo 82*). These are initially white but turn black as they mature.

The life cycle of *X. mali* is not clear. Only a few wild hosts have been reported, although *Xylaria* spp. are saprophytes on many plant species. Mycelium in dead wood in the soil is the most important inoculum source. The fungus can survive in infected roots for 10 years or longer. The role of ascospores and conidia in the disease cycle is unknown.

Control

Satisfactory controls are not available. Sites known to be infested with *Xylaria* should not be planted to susceptible plants. If replanting is attempted, all old roots should be carefully removed. There are some differences in rootstock susceptibility, although none are resistant. MM.104 and MM.111 are more susceptible than MM.106 and seedling rootstocks.

Selected References

Cooley, J. S. 1944. Some host-parasite relations in the black root rot of apple trees. *J. Agr. Res.*, 69:449–458.

Dozier, W. A., Jr., A. J. Latham, C. A. Kouskolekas and E. L. Manton. 1974. Susceptibility of certain apple rootstocks to black root rot and woolly apple aphids. *HortScience*, 9:35–36.

Fromme, F. D. 1928. *The black root rot disease of apple*. Tech. Bull. 34. Blacksburg, Va.: Virginia Agricultural Experiment Station.

Photo 82. Finger-like fruiting structures of the black root rot pathogen *Xylaria mali*.

Pome Fruits

SOUTHERN BLIGHT

Southern blight, caused by the fungus *Sclerotium rolfsii* Sacc., is primarily a problem in the Piedmont apple-growing areas in the southeastern United States. *Sclerotium rolfsii* is a widespread pathogen that affects several hundred plant species.

Symptoms and Disease Cycle

Sclerotium rolfsii affects the lower stems and roots of apple trees, killing the bark and girdling the trees (*Photo 83*). The disease is characterized by a white, web-like mycelium, which often forms at the bases and on the lower stems of affected trees (*Photo 84*). Light brown to yellow, round sclerotia ¹⁄₁₆ to ⅛ inch in diameter form in the mycelial mat. The fungus is spread by the sclerotia, which also serve as overwintering structures. The disease is most severe on 1- to 3-year-old trees. As the bark thickens, trees become resistant to infection.

Control

Avoid planting sites where the disease has been severe on previous crops such as peanut, clover, tomato and soybean. Keep the soil around the bases of trees free of dead organic matter that may serve as a food base for *S. rolfsii*. Some differences exist in rootstock susceptibility. The most resistant rootstock currently used is M.9. No fungicides are currently registered on apples for southern stem blight control.

Photo 83. Trees in a nursery row with dieback due to infection at soil line by the southern blight pathogen.

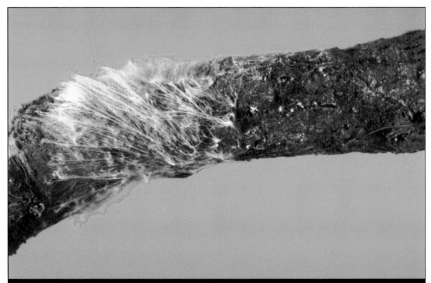

Photo 84. White, web-like mycelium of *Sclerotium rolfsii*. Light brown to yellow, round sclerotia often form in the mycelial mat.

Selected References

Brown, A. E., and F. F. Hendrix. 1980. Distribution and control of *Sclerotium rolfsii* on apple. *Plant Dis.*, 64:205–206.

Tomasino, S. F., and K. E. Conway. 1987. Spatial pattern, inoculum density-disease incidence relationship, and population dynamics of *Sclerotium rolfsii* on apple rootstock. *Plant Dis.*, 71:719–723.

Pome Fruits

APPLE UNION NECROSIS AND DECLINE

This disease was first reported on apple in Pennsylvania and Virginia, but subsequent observations and reports indicate it occurs in most apple-growing areas in the eastern United States, including Michigan. Certain clones of Delicious on MM.106 rootstock are the most severely affected.

Diseased trees generally exhibit symptoms of decline, including small, sparse, pale leaves; reduced terminal growth, and early discoloration of the foliage (*Photo 85*). Abnormally thick, spongy bark is found at the graft union. When it's removed, a distinct line of necrotic tissue may be observed at the graft union (*Photo 86*). Frequently, affected trees produce large numbers of sprouts from the rootstock, and sometimes the top breaks off at the union in strong winds (*Photo 87*).

This disease is caused by tomato ringspot virus, the same virus that causes Prunus stem pitting. For further information on this disease, including its control, see "Prunus stem pitting" in the Stone Fruit section of this bulletin.

Selected References

Rosenberger, D. A., M. B. Harrison and D. Gonsalves. 1983. Incidence of apple union necrosis and decline, tomato ringspot virus and *Xiphinema* vector species in Hudson Valley orchards. *Plant Dis.*, 67:356–360.

Rosenberger, D. A., J. N. Cummins and D. Gonsalves. 1989. Evidence that tomato ringspot virus causes apple union necrosis and decline: Symptom development in inoculated apple trees. *Plant Dis.*, 73:262–265.

Stouffer, R. F., K. D. Hickey and M. F. Welsh. 1977. Apple union necrosis and decline. *Plant Dis. Rep.*, 61:20–24.

Photo 85. Typical weak growth and pale coloration of foliage of Delicious on MM.106 rootstock with apple union necrosis and decline.

Photo 86. Line of dead tissue at the union between the Delicious scion and the MM.106 rootstock is typical of union necrosis and decline.

Photo 87. Breaking off of apple tree at the union between rootstock and scion is typical of union necrosis and decline.

Stone Fruits

AMERICAN BROWN ROT

Brown rot, caused by the fungus *Monilinia fructicola* (G. Wint.) Honey, is an economically important disease of apricot, peach, nectarine, plum and cherry. It reduces yields primarily by rotting fruit both on the tree and after harvest. In seasons with weather favorable for infection, entire crops may be lost, almost overnight.

Symptoms

Brown rot attacks blossoms, spurs, shoots and fruit. Fruit infections are the most destructive. Infected blossoms wilt, turn brown and persist into summer. Except on highly susceptible crops such as apricot, systemic infection of fruit-bearing spurs from infected blossoms is rare.

Outbreaks of brown rot are more common on mature than immature fruit. Initially, small, circular, light brown spots develop on the surface of the fruit and expand rapidly under favorable conditions, destroying entire fruit (Photos 88, 89). Rotted fruit may fall to the ground or persist as mummies on the tree.

Spurs and small branches, particularly of peach, nectarine and apricot, may wither and die because of the fungus growing into them from infected fruit and mummies (Photo 89). Gummosis often accompanies the blighting of spurs and formation of cankers. Succulent shoots are sometimes blighted from infection initiated through wounds.

Decay of apple and pear fruit by *M. fructicola* is rare; then it

occurs only when fruit become overripe in the autumn or when cracking or other injury exposes the flesh to infection (Photo 90).

Under wet and humid conditions, ash-gray tufts (sporodochia) bearing conidia of the fungus develop over the surface of the infected tissues (Photo 91). The presence of conidia on lesions is the most distinctive characteristic of brown rot.

Photo 88. Decay of sour cherries by *Monilinia fructicola* and sporulation of the pathogen.

Photo 89. Decay of peach and dieback of the shoot growth from brown rot.
(Courtesy M. A. Ellis, Ohio Agricultural Research and Development Center, Wooster)

Photo 90. Brown rot on apple.
(Courtesy M. A. Ellis, Ohio Agricultural Research and Development Center, Wooster)

Photo 91. Sporulation of the brown rot fungus on peach following infection of the shoot tip.

Disease Cycle

Sources of inoculum for blossom blight are overwintering brown rot–infected mummies, peduncles, cankers (*Fig. 5*) and, infrequently, apothecia (saucer-like, sexual, ascus–bearing fruiting bodies) from mummies on the orchard floor (*Photo 92*). Overwintering of the brown rot fungus in twig cankers is more common on peach, nectarine and apricot than on cherry. Conidia from mummies and other sources are disseminated by splashing or wind–driven rain, and ascospores discharged from apothecia during rain are carried by wind to blossoms. Infection of the blossoms depends on duration of wetness and temperature. The hours of wetting necessary for blossom infection decrease from 18 hours at 50 degrees F to 5 hours at 77 degrees F. Infection proceeds slowly above 80 and below 55 degrees (minimum 40 degrees F).

The potential for brown rot infection increases as the fruit mature. Infection occurs directly through the cuticle, through natural openings in fruit and through wounds. Injured fruit, such as sweet cherries cracked from rain, are particularly susceptible to infection. Brown rot mummies, infected blossoms, nonabscised aborted peach fruit, thinned fruit on the orchard floor and fruit infected during the growing season are inoculum sources for fruit infection. The level of inoculum is an important factor governing the severity of brown rot outbreaks on fruit. Warm, wet, humid weather lasting 2 to 3 days favors sporulation; long, dry periods inhibit sporulation. Decay of the fruit and subsequent conidial production may occur in a few days, so the disease can build up rapidly.

Control

Brown rot is controlled with a combination of sanitation practices to reduce the amount of fungal inoculum and a protective fungicide program. Because of the small number of disease–resistant cultivars, host resistance is not a major component of disease management programs. Removing fruit, mummies and blighted twigs from trees after the final harvest reduces the amount of brown rot inoculum at the beginning of the next season. Cultivation just before bloom and no later than midbloom will destroy the apothecia by disturbing the mummies.

Blossom infections are controlled with two to three fungicide sprays during the bloom period. The number of spray applications for brown rot required during bloom varies from year to year, depending on the weather, the susceptibility of the species of stone fruit, the

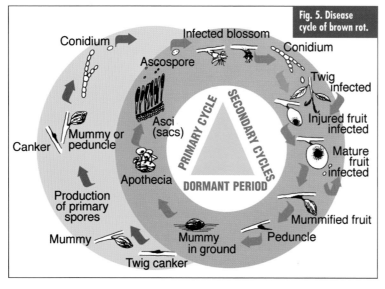

Fig. 5. Disease cycle of brown rot.

Photo 92. Apothecia or cup-shaped fruiting bodies of brown rot arising from a peach mummy.

length of the bloom period and the type of fungicide. Generally, one or two sprays are applied, but none may be needed some years on less susceptible species such as sour cherry, while in other years, three or four sprays may be needed on highly susceptible species such as apricot. Sweet cherry is generally more susceptible to brown rot than tart cherry, and nectarines are more susceptible than peaches.

Brown rot is controlled on ripening fruit with two to three preharvest fungicide treatments initiated when the fruit begins to color (generally about 3 weeks before harvest). Controlling insects, particularly those pests that directly injure fruit, helps to prevent infection.

Picking and handling fruit carefully to avoid injuries, removing field heat from the fruit promptly after harvest by hydrocooling or forced air cooling, using clean containers, and periodically removing ripe and rotting fruit from packing areas all help to prevent losses from decay during storage and in transit. Other techniques that reduce decay are hot water treatments, fungicidal dips, wax-fungicide treatments and fungicide sprayed on fruit during grading.

Selected References

Biggs, A. R., and J. Northover. 1985. Inoculum sources of *Monilinia fructicola* in Ontario peach orchards. *Can. J. Plant Pathol.*, 7:302–307.

Biggs, A. R., and J. Northover. 1988. Influence of temperature and wetness duration on infection of peach and sweet cherry fruits by *Monilinia fructicola*. *Phytopathology*, 78:1352–1356.

Biggs, A. R., and J. Northover. 1988. Early and late-season susceptibility of peach fruits to *Monilinia fructicola*. *Plant Dis.*, 72:1070–1074.

Biggs, A. R., and J. Northover. 1990. Susceptibility of immature and mature sweet and sour cherries to *Monilinia fructicola*. *Plant Dis.*, 74:280–284.

Byrde, R. S. W., and H. S. Willetts. 1977. *The Brown Rot Fungi of Fruit. Their Biology and Control.* Oxford: Pergamon Press.

Sutton, T. B., and C. N. Clayton. 1972. Role and survival of *Monilinia fructicola* in blighted peach branches. *Phytopathology*, 62:1369–1373.

Wilcox, W. F. 1989. Influence of environment and inoculum density on the incidence of brown rot blossom blight of sour cherry. *Phytopathology*, 79:530–534.

Wilcox, W. F. 1990. Eradicant and antisporulant activities of dicarboximide and sterol demethylation inhibitor fungicides in control of blossom blight caused by *Monilinia fructicola*. *Plant Dis.*, 74:808–811.

Stone Fruits

EUROPEAN BROWN ROT

European brown rot, caused by *Monilinia laxa* (Aderh. and Ruhl.) Honey, is endemic on sour cherry in Michigan, New York and Wisconsin. The disease is rare on Montmorency, but it can be a significant problem on the cultivars Meteor, English Morello and Balatan (Érdi bőtermő).

Symptoms and Disease Cycle

The fungus infects and kills blossoms and spurs (*Photos 93, 94*).

Photo 93. Spur dieback of Montmorency sour cherry caused by the European brown rot fungus, *Monilinia laxa*.

Infections on highly susceptible sour cherry cultivars are distributed throughout the orchard. Infections on Montmorency sour cherry are usually found along hedgerows or in low areas of the orchard where blossoms tend to dry slowly. Wet periods lasting for a day or more are required for severe blossom infection.

Newly infected blossoms turn brown and the fungus sporulates profusely on them. Leaves at the base of the blossom also are invaded and killed, followed by systemic infection of the spur (*Photo 95*). One- to 3-inch-long elliptical cankers are formed at the bases of blighted spurs. In subsequent seasons, conidia are produced on blossom debris, dead spurs and cankers if adequate moisture is present when trees come into bloom. Fruit infections are rare, although fruit in direct contact with infected blossoms may become infected.

Control

The common sweet cherry cultivars grown in the northeastern and north central United States are resistant to *M. laxa*. Montmorency sour cherry has sufficient resistance that fungicide sprays are very rarely required. Removing hedgerows that create a microclimate favorable for disease development is often adequate for preventing the disease on Montmorency.

On highly susceptible cultivars, the disease can be controlled with two fungicide sprays. The first spray is applied at the popcorn stage and the second about 7 days later. Pruning out old infections during the winter helps to reduce inoculum levels at bloom.

Selected References

Calavan, E. C., and G. W. Keitt. 1948. Blossom and spur blight (*Sclerotinia laxa*) of sour cherry. *Phytopathology*, 38:857–882.

Kable, P. F., and K. G. Parker. 1963. The occurrence of the imperfect stage of *Monilinia laxa* on *Prunus cerasus* var. *austera* in New York state. *Plant Dis. Rep.*, 47:1104.

Tamm, L., and W. Flückiger. 1993. Influence of temperature and moisture on growth, spore production and conidial germination of *Monilinia laxa*. *Phytopathology*, 83:1321–1326.

Photo 95. Canker and gumming around a fruiting spur killed by European brown rot. Bark was removed to reveal canker.

Photo 94. Spur dieback of Meteor sour cherry caused by European brown rot.

Stone Fruits

CHERRY LEAF SPOT

Cherry leaf spot, caused by the fungus *Blumeriella jaapii* (Rehm) Arx, is an important disease of sour and sweet cherries in the Great Lakes and mid-Atlantic states. After a severe outbreak of leaf spot in Michigan in 1993, thousands of sour cherry trees died during the following winter. The disease is of minor importance on prunes. The name "shot-hole," although sometimes used when referring to cherry leaf spot, should not be used because of possible confusion with a different shot-hole disease of stone fruits that occurs in the western United States.

Symptoms

Leaf spot is primarily a disease of the foliage and consequently can often adversely affect the vigor and health of trees. Lesions first appear on the upper surfaces of leaves as small reddish to purple spots (*Photo 96*). These spots turn brown and may coalesce if numerous. On plums, and occasionally on cherries, the necrotic circular lesions drop out, producing shot-hole symptoms. During rainy weather, light pink to white masses of conidia appear on the underside of the leaf in the centers of the spots (*Photo 97*). Only a few lesions per leaf can result in yellow or chlorotic leaves. Chlorotic infected leaves abscise, often resulting in severe defoliation (*Photo 98*). Infection of fruit and fruit pedicels is rare but may be observed in severe epidemics or on some cultivars (*Photo 99*).

Fruit on trees severely defoliated before harvest fail to mature normally and are light-colored, low in soluble solids, soft and watery. Flower bud formation and fruit set on severely defoliated trees may be reduced for at least two seasons. Trees defoliated in mid-summer, particularly young non-bearing trees, are less cold hardy and may be killed by low temperatures in winter (*Photo 100*).

Photo 96. Cherry leaf spot infection on sweet cherry leaves.

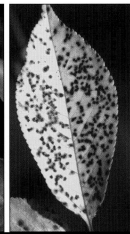

Photo 97. Sporulation of the pathogen in acervuli on the lower surface (left) and small, circular spots plus yellowing of sour cherry leaves (right) caused by cherry leaf spot.

Photo 98. Defoliation by cherry leaf spot of unsprayed trees (right) compared with no defoliation on fungicide-sprayed tree (left).

Photo 99. Stems of sour cherry fruit with sporulating cherry leaf spot lesions. Note single lesions on the two right fruit.

Photo 100. Death of cherry tree on left in winter following severe defoliation from leaf spot the previous summer. Leaf spot was controlled on the tree that is in bloom.

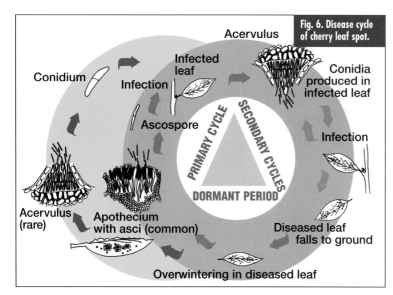

Fig. 6. Disease cycle of cherry leaf spot.

Acervulus

Conidium

Infection

Infected leaf

Conidia produced in infected leaf

Ascospore

PRIMARY CYCLE · SECONDARY CYCLES

DORMANT PERIOD

Infection

Acervulus (rare)

Apothecium with asci (common)

Diseased leaf falls to ground

Overwintering in diseased leaf

Disease Cycle

The fungus overwinters in old infected leaves on the orchard floor (*Fig. 6*) and in the spring produces apothecia or open, saucer–like, ascus-bearing fruiting bodies (*Photos 101, 102*). The optimum temperature for apothecial development is 62 degrees F. Ascospores may be discharged during and shortly after rainfall from early bloom to about 6 weeks after petal fall. Ascospore discharge is highest at 60 to 85 degrees F and lowest at 41 to 46 degrees F. Low numbers of acervuli or saucer–like conidia-bearing fruiting bodies may also be produced in old infected leaves.

Infection takes place through stomates (air pores) located on the leaf undersides (*Photo 103*). Once unfolded, leaves are susceptible throughout the season, but susceptibility decreases with age. Sweet cherries are less susceptible to leaf spot than sour cherries.

The partial resistance of sweet cherries is reflected in longer latent periods, lower rates of lesion expansion and the production of fewer conidia per lesion.

Infection by ascospores and conidia is governed by the duration of wetting from rain relative to temperature (*Table 2, page 59*). A wet period of only a few hours is sufficient for spore germination and infection when temperatures are favorable. Development of visible lesions occurs in 5 to 15 days, depending on temperature and moisture conditions. Optimum conditions for lesion development are temperatures of 60 to 68 degrees F with rainfall or high humidity. As lesions appear, acervuli containing masses of whitish pink conidia are visible on the leaf undersurfaces. The conidia are disseminated from leaf to leaf by splashing rain and wind. Secondary spread and infection by conidia continues in repeated cycles until autumn leaf fall.

Control

All commercially acceptable cultivars of cherry are susceptible to leaf spot, so the primary approach to control is the use of fungicide sprays. Applications

Photo 101. Apothecia of the leaf spot fungus on overwintered sweet cherry leaves.

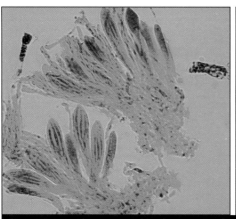
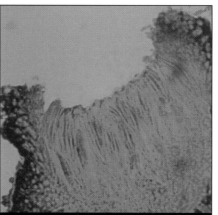

Photo 102. Asci with ascospores (left) and cross-section of an apothecium (right) of the cherry leaf spot fungus, *Blumeriella jaapii*.

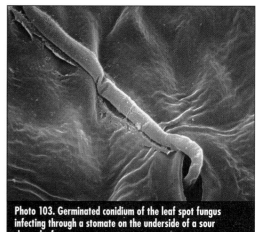

Photo 103. Germinated conidium of the leaf spot fungus infecting through a stomate on the underside of a sour cherry leaf.

Table 2. Approximate number of hours of wetting period required for conidial infection by the cherry leaf spot fungus at various air temperatures.[a]

Average temperature		Wetting period (hr)[b]		
(°F)	(°C)	Light infection	Moderate infection	Heavy infection
81	27.2	28	43	—
80	26.7	21	35	—
79	26.1	18	30	—
78	25.5	16	27	42
77	25.0	14	24	36
76	24.4	12	21	32
75	23.8	11	19	29
74	23.3	9	18	27
73	22.7	8	16	25
72	22.2	7	15	23
71	21.6	7	14	22
70	21.1	6	13	21
69	20.5	6	13	20
63–68	17.2–20.0	5	12	19
62	16.6	6	12	19
61	16.1	6	13	20
60	15.5	7	13	20
59	15.0	7	14	21
58	14.4	8	15	22
57	13.8	9	16	23
56	13.3	10	17	24
55	12.7	11	18	25
54	12.2	12	19	27
53	11.6	14	21	29
52	11.1	15	23	31
51	10.5	17	25	33
50	10.0	19	27	35
49	9.4	20	29	38
48	8.8	23	32	42
47	8.3	25	34	46
46	7.7	28	38	51

[a]Adapted from Eisensmith and Jones, 1981. *Plant Dis.,* 65:955–958, and *Phytopathology,* 71:728–732.

[b]The infection period is considered to start when rain begins.

are started at petal fall or after the first leaves have unfolded, repeated every 7 to 10 days to harvest, and concluded with one or two postharvest applications beginning 2 to 3 weeks after harvest. Efficiency of fungicide use can be improved by spraying alternate sides of trees on a 7-day schedule rather than spraying each side on a 10-day schedule. The disease is more difficult to control on sour cherries than on sweet cherries or European plums because of their high susceptibility.

Selected References

Eisensmith, S. P., and A. L. Jones. 1981. A model for detecting infection periods of *Coccomyces hiemalis* on sour cherry. *Phytopathology,* 71:728–732.

Eisensmith, S. P., A. L. Jones and C. E. Cress. 1982. Effect of interrupted wet periods on infection of sour cherry by *Coccomyces hiemalis. Phytopathology,* 72:680–682.

Eisensmith, S. P., T. M. Sjulin, A. L. Jones and C. E. Cress. 1982. Effects of leaf age and inoculum concentration on infection of sour cherry by *Coccomyces hiemalis. Phytopathology,* 72:574–577.

Garcia, S. M., and A. L. Jones. 1993. Influence of temperature on apothecial development and ascospore discharge by *Blumeriella jaapii. Plant Dis.,* 77:776–779.

Howell, G. S., and S. S. Stackhouse. 1973. The effect of defoliation time on acclimation and dehardening in tart cherry (*Prunus cerasus* L.). *J. Am. Soc. Hortic. Sci.,* 98:132–136.

Jones, A. L., G. R. Ehret, S. M. Garcia, C. D. Kesner and W. M. Klein. 1993. Control of cherry leaf spot and powdery mildew on sour cherry with alternate-side applications of fenarimol, myclobutanil and tebuconazole. *Plant Dis.,* 77:703–706.

Keitt, G. W., E. G. Blodgett, E. E. Wilson and R. O. Magie. 1937. *The epidemiology and control of cherry leaf spot.* Res. Bull. 132. Madison, Wis.: Wisconsin Agricultural Experiment Station.

Sjulin, T. M., A. L. Jones and R. L. Andersen. 1989. Expression of partial resistance to cherry leaf spot in cultivars of sweet, sour, duke and European ground cherry. *Plant Dis.,* 73:56–61.

PEACH SCAB

Peach scab, caused by the fungus *Cladosporium carpophilum* Thüm., also attacks apricot and nectarine. It is of minor economic importance on peach in the upper Midwest but of major economic importance in the southeastern United States. It reduces the appearance, quality and market value of the fruit, and when severe, it creates entry points for the brown rot pathogen and other decay-causing fungi.

Symptoms

Peach scab develops as circular, olivaceous to black, velvety spots on fruit, twigs and leaves. Fruit lesions tend to be concentrated at the stem end of half-grown to mature fruit (*Photos 104, 105*). Lesions on fruit coalesce when numerous and may be followed by cracking of the fruit. Shoot and twig infections are slightly raised and circular to oval, becoming brown with slightly raised purple margins later in the season (*Photo 106*).

Disease Cycle

The fungus overwinters in twig lesions. Conidial production in these lesions normally begins at shuck split and peaks 2 to 6 weeks later. Maximum sporulation occurs after 24 hours of high relative humidity. Conidia are airborne and waterborne. Fruit are susceptible to infection from the shuck fall stage until harvest, and new infections remain latent for 40 to 60 days. Conidia produced in lesions on fruit and twigs also infect young shoots and leaves.

Control

Disease control is achieved by pruning and the use of protectant fungicides. Pruning helps increase air circulation, facilitates drying of fruit and foliage, and improves spray penetration into the trees. Fungicide sprays, applied at 10- to 14-day intervals, should begin at the shuck split stage in areas where the disease is a serious problem, and about 10 days later where it is minor. Fungicide protection is especially critical 2 to 6 weeks after shuck split, and protection should be maintained until about 40 days before harvest.

To avoid introducing the disease into new plantings via infected nursery trees, purchase peach trees that have been sprayed routinely with a fungicide at the nursery.

Photo 104. Peach with brown to black, velvety lesions caused by the scab fungus *Cladosporium carpophilum*.

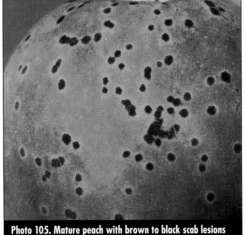

Photo 105. Mature peach with brown to black scab lesions with yellow margins.

Photo 106. Black, depressed lesions on a peach twig caused by the scab fungus *Cladosporium carpophilum*.

Selected References

Gottwald, T. R. 1983. Factors affecting spore liberation by *Cladosporium carpophilum*. *Phytopathology, 73*:1500–1505.

Keitt, G. W. 1917. *Peach scab and its control*. Bull. 395. Washington, D.C.: U.S. Department of Agriculture.

Lawrence, E. G., Jr. and E. I. Zehr. 1982. Environmental effects on the development and dissemination of *Cladosporium carpophilum* on peach. *Phytopathology, 72*:773–776.

Stone Fruits

BACTERIAL SPOT

Bacterial spot, originally described on Japanese plums from Michigan in 1902, is caused by the bacterium *Xanthomonas campestris* pv. *pruni* (Smith) Dye. The disease is a serious problem on apricot, peach, nectarine, prune and plum throughout the eastern United States. Highly susceptible cultivars cannot be grown profitably in this region because entire crops can be lost in years with extended periods of warm, wet, humid weather during the growing season.

Symptoms

Leaves, fruit and shoots may exhibit disease symptoms. Leaf lesions are small (1 to several millimeters in diameter) and generally angular. Initially, lesions appear as water-soaked, grayish areas primarily on the undersides of leaves, later as brown to purplish black spots. The centers of the spots may fall out, giving the leaves a tattered appearance. Lesions are frequently concentrated along midribs and at the tips of leaves, where the tissue may become necrotic as lesions coalesce (*Photo 107*).

Severely infected leaves turn yellow and drop. On sensitive cultivars, a few lesions result in severe defoliation; tolerant cultivars require many more. Severe defoliation often results in reduced fruit size, increased sunburn and cracking of the fruit, and reduced tree vigor and winter hardiness.

Fruit infected early in the growing season develop unsightly blemishes in the skin (*Photos*

Photo 107. Yellowing and tip burning of peach leaves with bacterial spot, caused by *Xanthomonas campestris* pv. *pruni*.

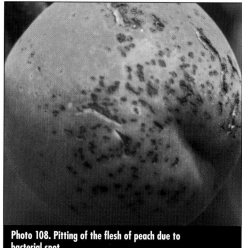

Photo 108. Pitting of the flesh of peach due to bacterial spot.

Photo 109. Bacterial spot on apricot.

Photo 110. Stanley plum with bacterial spot. Note purple lesions.

108–111). Pits or cracks extend into the flesh, resulting in depressed, brown to black lesions that often coalesce to affect large areas of the fruit. Lesions on young fruit may exhibit gumming; those that develop on fruit during the preharvest period tend to be superficial, giving the fruit a mottled appearance.

Infection of current season's growth can result in the production of elliptical cankers in the summer or the following spring. Lesions that develop in summer on green shoots and twigs are called summer cankers. Lesions that develop after bud break in

Photo 111. Pitting and gumming on nectarines with bacterial spot.

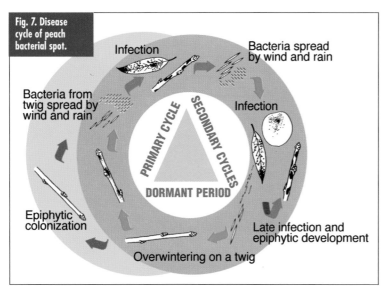

Fig. 7. Disease cycle of peach bacterial spot.

PRIMARY CYCLE — SECONDARY CYCLES — DORMANT PERIOD

Infection

Bacteria spread by wind and rain

Infection

Bacteria from twig spread by wind and rain

Late infection and epiphytic development

Epiphytic colonization

Overwintering on a twig

Photo 112. Bacterial spot cankers on peach at bud break in spring.

the spring on the previous season's growth are called spring cankers (*Photo 112*). Summer cankers are usually located between the nodes; spring cankers tend to be located at nodes.

Disease Cycle

The pathogen overwinters in twig lesions and buds and on symptomless plant surfaces (*Fig. 7*). In spring, the bacteria are spread by windblown rain to leaves, fruits and shoots. Moisture from fog and dew tends to wash bacteria toward tips of leaves. Water congestion of tissues is important for infection to occur

and for disease development. Therefore, hard, driving rains are more important than gentle rains in initiating new infections. This explains why symptoms are often more severe on the side of trees exposed to prevailing winds.

Secondary spread of bacteria oozing from summer cankers and leaf and fruit lesions can occur in repeated cycles during warm, wet weather. Driving rain, high humidity, moderate temperatures and high winds favor infection; hot, dry weather inhibits bacterial spread and disease development. Systemic movement of bacteria from infected leaves and shoots contributes to the formation of cankers and to spread by budding to healthy nursery trees.

Control

Disease control is achieved by avoiding susceptible cultivars. Locating new plantings of peach, apricot, prune and nectarine in close proximity to orchards with susceptible cultivars contributes to a buildup of disease.

Some of the more susceptible peach cultivars are Blake, Jersey-land, Suncrest, Suncling, Sunhigh and Jersey Queen. J.H. Hale, Babygold 5, Elberta, Kalhaven and Rio-Oso-Gem are moderately susceptible. Relatively resistant cultivars are Redskin, Redhaven, Loring, Candor, Biscoe, Dixired, Sunhaven, Jefferson and Madison. Some recently released cultivars that look good are Salem, Contender, Harrow Beauty

and Harrow Diamond. Vulcan and Vinegold, new clingstone peach cultivars from the Vineland Horticultural Experiment Station of Ontario, may be of interest to commercial processing peach growers because of their resistance to bacterial spot.

Chemical sprays help to reduce the amount of fruit and leaf infection. For optimal control, they must be applied before symptoms occur. The first spray is usually a copper compound applied just before tree growth resumes in the spring. It is followed by frequent applications of an antibiotic beginning at petal fall. The 3-week period following petal fall is critical for early-season fruit infection and establishment of inoculum on new foliage. Rainfall during this period is important for infection. These programs, however, only suppress development of the disease–they do not eliminate it. Because of the uncertainty of chemical control, the best control strategy for bacterial spot is the use of resistant cultivars.

Selected References

Feliciano, A., and R. H. Daines. 1970. Factors influencing ingress of *Xanthomonas pruni* through peach leaf scars and subsequent development of spring cankers. *Phytopathology,* 60:1720–1726.

Miles, W. G., R. H. Daines and J. W. Rue. 1977. Presymptomatic egress of *Xanthomonas pruni* from infected peach leaves. *Phytopathology,* 67:895–897.

Shepard, D. P., and E. I. Zehr. 1994. Epiphytic persistence of *Xanthomonas campestris* pv. *pruni* on peach and plum. *Plant Dis.,* 78:627–629.

Stone Fruits

Peach Leaf Curl

Peach leaf curl, caused by the fungus *Taphrina deformans* (Berk.) Tul., is a common disease of peach and nectarine throughout the eastern United States. Tree vigor, fruit quality and yield can be reduced by defoliation early in the growing season.

Photo 113. Deformed peach leaves with leaf curl.

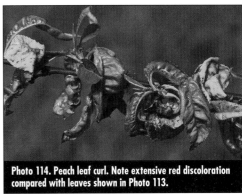

Photo 114. Peach leaf curl. Note extensive red discoloration compared with leaves shown in Photo 113.

Symptoms

Intercellular growth of the fungus inside the current season's growth causes cell division and cell enlargement. This results in thickened, curled to convoluted or blistered leaves, shoots and fruit. Infected leaves, first evident about a month after full bloom, are often flushed with red or purple (*Photos 113, 114*). Symptoms may be limited to small areas of the leaf and to a few leaves, or they may involve the entire leaf and most leaves on a tree. Infected leaves eventually become necrotic, wither and fall, resulting in defoliation.

The growth of infected shoots is stunted, swollen, chlorotic and rosetted. Infected flowers are distorted and usually drop before symptoms are well developed. Infected fruit exhibit raised, wrinkled areas with reddish discoloration and often fall prematurely (*Photo 115*).

Disease Cycle

Soon after symptoms are visible, a powdery–gray, felt–like, ascus–bearing fungal growth breaks through the cuticle of infected tissues (*Photo 116*). Each ascus contains ovoid ascospores, which are released when the mature asci rupture. During periods of wet weather, the asco-

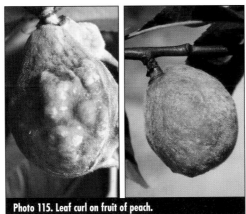

Photo 115. Leaf curl on fruit of peach.

spores produce large numbers of blastoconidia or budconidia by budding.

Infection occurs at bud burst when ascospores and budconidia are disseminated in water from overwintering sites on the bark to buds with loose bud scales. Prolonged periods of cool, wet, rainy weather at bud burst favor severe infection. Temperatures between 50 and 70 degrees F are optimum for infection.

Control

In most areas of the eastern United States, the disease can be prevented by applying a single fungicide spray in the autumn or before bud swell in spring. No cultivar is immune to leaf curl. Redhaven and most cultivars derived from Redhaven are more resistant to leaf curl than Redskin and cultivars derived from Redskin.

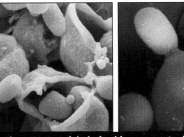

Photo 116. Asci of the leaf curl fungus emerging from the surface of a leaf (top). Ruptured ascus containing ascospores and producing a bud conidium (bottom, left). An ascospore producing a bud conidium (bottom, right).

To help maintain tree vigor where leaf curl is severe, thin fruit more than normal, reduce drought stress by periodic irrigation and apply an extra application of nitrogen fertilizer.

Selected References

Fitzpatrick, R. E. 1934. The life history and parasitism of *Taphrina deformans*. *Sci. Agr.*, 14:305–326.

Mix, A. J. 1935. The life history of *Taphrina deformans*. *Phytopathology*, 25:41–66.

Ritchie, D. F., and D. J. Werner. 1981. Susceptibility and inheritance of susceptibility to peach leaf curl in peach and nectarine cultivars. *Plant Dis.*, 65:731–734.

Stone Fruits

PLUM POCKETS

Plum pockets, caused by *Taphrina pruni* Tul., occurs on wild or abandoned plum trees–it is rare and of no economic importance in plum orchards. It is recognized by the production of abnormally large, misshapen, bladder–like fruit with thick, spongy flesh (*Photo 117*). Infected fruit are hollow in the center because of the abortion of the seed (*Photo 118*). The surface of the fruit is often covered with a powdery–gray, ascus–bearing fungal growth. Infected fruit become brown, wither and eventually fall from the tree.

A spray program similar to the one for peach leaf curl may control plum pockets as well.

Selected References

Hickey, K. D. 1995. Plum pockets. Pages 19–20 in: *Compendium of Stone Fruit Diseases.* J. M. Ogewa, E. I. Zehr, G. W. Bird, D. F. Ritchie, K. Uriu and J. K. Uyemoto, eds. St. Paul, Minn.: American Phytopathological Society.

Photo 117. Plum pockets. Note enlarged infected fruit on left and normal-sized healthy fruit on right.

Photo 118. Hollow interior of plum pockets–infected plums.

Stone Fruits

Black Knot of Plum

Black knot, caused by the fungus *Apiosporina morbosa* (Schwein.:Fr.) Arx, is a serious disease of plums and prunes in eastern North America and of sour cherry in southern Ontario, Canada. It is also destructive on wild plum and wild cherry trees. The economic value of plum and prune orchards may be destroyed in 2 to 3 years if no attempt is made to control the disease.

Symptoms

Black knot is characterized by the production of elongated, corky outgrowths or knots on shoots, spurs, branches and trunks (*Photos 119, 120*). The knots tend to be longer than they are wide, occasionally reaching a foot or more in length on large branches and trunks. Newly formed knots have greenish, soft tissue; old knots have black, hard tissue that is often riddled by insects and parasitized by a white or pink fungus. Limbs and sometimes whole trees are stunted and eventually killed by the girdling action of the ever-expanding knots.

Disease Cycle

Ascospores are the only proven infection propagule of the pathogen. They infect the host through succulent vegetative shoots from bud break until terminal growth stops. Conidia produced on the surface of knots are not believed to cause infection.

In spring, at the green-tip stage of bud development, ascospore discharge is initiated from perithecia located in 2-year-old knot tissue. Rain is required for ascospore discharge, and maximum discharge usually occurs between pink and 2 weeks after bloom. The ascospores are disseminated by wind and by splashing rain. Free moisture is required for the initiation of infection, and temperatures between 55 and 77 degrees F favor disease development.

Several months are required between infection and the appearance of knots. Some knots are visible in the autumn of the year infection occurred; others are visible in late spring and in early summer the following year. Perithecia are usually produced in 2-year-old knot tissue. The fungus remains active in the tissues of the host throughout the growing season, extending the knots an inch or more each year.

Cultivar Susceptibility

The three major plum cultivars grown in eastern North America—Stanley, Bluefre and Damson—are susceptible to black knot. Cultivar susceptibility tests in Pennsylvania show that Shropshire and Stanley are highly susceptible; Methley, Milton, Early Italian, Brodshaw and Fellenburg are moderately susceptible; Shiro, Santa Rosa and Formosa are slightly susceptible, and President apparently is resistant to black knot.

Control

Control of black knot is based on a combination of cultural and

Photo 119. Young (top) and old (bottom) knots on plum shoots caused by black knot.

Photo 120. Black knot lesion on plum after about 3 years.

chemical control methods. Infected wild plum and cherry seedlings should be removed from fencerows and woodlots along orchard perimeters. Adjacent plum and cherry orchards with black knot–infected trees should also be removed. Inspect orchards and surrounding wooded areas each winter for knots and prune out infected shoots and limbs. The fungus may have extended beyond the visible swelling, so make cuts well below (2 to 3 inches) the margin of each knot. Remove knots from the orchard and burn them before bud break–they may remain a source of inoculum if left on the orchard floor.

Very few fungicides are efficacious against black knot, and fungicides effective in one area are sometimes not effective in other areas. Fungicides, applied from white bud through shuck split (green tip through second cover in problem orchards) in conjunction with cultural control, will help prevent the disease. Spraying alone may not give adequate control.

Selected References

Northover, J., and W. McFadden-Smith. 1995. Control and epidemiology of *Apiosporina morbosa* of plum and sour cherry. *Can. J. Plant Pathol.*, 17:57–68.

Ritchie, D. F., E. J. Klos and K. S. Yoder. 1975. Epidemiology of black knot of "Stanley" plums and its control with systemic fungicides. *Plant Dis. Rep.*, 59:499–503.

Rosenberger, D. A., and W. D. Gerling. 1984. Effect of black knot incidence on yield of Stanley prune trees and economic benefits of fungicide protection. *Plant Dis.*, 68:1060–1064.

Smith, D. H., F. H. Lewis and S. H. Wainwright. 1970. Epidemiology of the black knot disease of plums. *Phytopathology*, 60:1441–1444.

Wainwright, S. H., and F. H. Lewis. 1970. Developmental morphology of the black knot pathogen on plum. *Phytopathology*, 60:1238–1244.

Stone Fruits

POWDERY MILDEW OF CHERRY

Powdery mildew, caused by *Podosphaera clandestina* (Wallr.:Fr.) Lév., can reduce the growth of sour cherry trees in the nursery and young orchards. A high incidence of mildew on leaves of bearing trees can result in increased leaf removal from mechanical harvesting. Mildew is rare on sweet cherry in the eastern United States.

Symptoms and Disease Cycle

Mildew on young leaves appears first as circular, white, felt–like patches of fungal mycelium and conidia. Lesions spread rapidly, often engulfing the entire leaf (*Photo 121*). Numerous small, brown to black, spherical bodies (cleistothecia) with dichotomously branched appendages develop in the felt–like patches as the season progresses (*Photo 122*). Severely infected leaves exhibit upward rolling, becoming stiff and brittle with age. Mildew on green fruit appears as shiny, red blotches, often with white mycelium and spores in the center (*Photo 123*). Fruit infections are very rare in the eastern United States.

Fungal spores (conidia) produced in developing infections are carried by wind to young leaves, where they initiate secondary infections. The optimum temperature for conidial germination is 68 degrees F. Growth of the fungus across the leaf surface is supported by haustoria, specialized outgrowths of the fungus that penetrate the epidermal leaf cells and extract nutrients.

Photo 121. White fungal growth of the powdery mildew fungus on leaves of sour cherry.

Photo 122. Cleistothecium of cherry mildew with dichotomous branching of appendages.

Mildew is favored by dry summers with intermittent periods of high humidity and moisture.

Beginning in midsummer, cleistothecia develop in the felt–like patches. Each cleistothecium contains a single ascus with eight ascospores. The fungus overwinters as cleistothecia in fallen leaves. They release ascospores the following spring to initiate primary infections.

Control

Where powdery mildew is an economic problem in sour cherry orchards, the disease can be prevented with fungicides. Fungicide spray programs should begin at petal fall or shuck split and be continued at 7- to 10–day intervals until harvest. Such cultural practices as pruning orchard trees and removing hedgerows located very close to orchards facilitate drying of foliage and fruit and create a less favorable microclimate for disease development.

Photo 123. Red blotches on Montmorency sour cherry caused by powdery mildew.

Selected References

Grove, G. G., and R. J. Boal. 1991. Overwinter survival of *Podosphaera clandestina* in eastern Washington. *Phytopathology,* 81:385–391.

Grove, G. G., and R. J. Boal. 1991. Factors affecting germination of conidia of *Podosphaera clandestina* on leaves and fruit of sweet cherry. *Phytopathology,* 81:1513–1518.

Jones, A. L., G. R. Ehret, S. M. Garcia, C. D. Kesner and W. M. Klein. 1993. Control of cherry leaf spot and powdery mildew on sour cherry with alternate-side applications of fenarimol, myclobutanil and tebuconazole. *Plant Dis.,* 77:703–706.

Stone Fruits

Powdery mildew, caused by the fungus *Sphaerotheca pannosa* (Wallr.:Fr.) Lév., and rusty spot, a disease associated with mildew fungi, are rare in peach orchards in Michigan. They occur sporadically and are of minor economic importance on peach in the mid–Atlantic and southeastern United States. Fruit infections (rusty spot) cause the greatest economic loss; infections to leaves and shoots can reduce vegetative growth, particularly of nursery trees.

Symptoms

Leaves and shoots of nursery stock and, rarely, of young, vigorously growing orchard trees are covered with white, felt–like mycelium and later become distorted and stunted. White, circular spots on young fruit expand in size. Later the mycelium sloughs off, leaving a rusty–colored patch with dead epidermal cells (*Photos 124, 125*). The rusty spot expands as fruit enlarge.

Disease Cycle

The mildew fungus overwinters as mycelium in dormant peach buds. When infected shoots emerge in spring, the pathogen initiates its growth and produces conidia. Conidia are carried by wind and rain to leaves, twigs and fruit. Infections may also be initiated by conidia blown in from sources outside the planting. Other stone fruit crops and roses are common external sources of inoculum.

Leaves are susceptible to infection when young but become resistant as they mature. Fruit are also susceptible to infection when young and become resistant at the pit hardening stage. The disease is most severe following periods of moderate temperatures and high humidity. Spore germination can occur at relative humidities of 43 to 100 percent. Free moisture is not required. The optimum temperature for germination is 66 to 72 degrees F. Under favorable environmental conditions, new lesions develop in 10 days.

Control

Powdery mildew control is achieved by avoiding susceptible cultivars and using fungicides. Most peach cultivars are resistant to powdery mildew. Rio–Oso–Gem and Redskin are susceptible and may need spray treatments. Fungicide sprays should be initiated at petal fall and continued every 10 to 14 days until the pit hardening stage of fruit development. Additional sprays may be needed until growth stops to prevent infection of leaves and shoots.

Selected References

Ries, S. M., and D. J. Royse. 1978. Peach rusty spot epidemiology: Incidence as affected by distance from a powdery mildew-infected apple orchard. *Phytopathology,* 68:896–899.

Weinhold, A. R. 1961. The orchard development of peach powdery mildew. *Phytopathology,* 51:478–481.

Photo 124. White fungal growth of the powdery mildew fungus on peach.

Photo 125. Reddish (left) and rusty-colored (right) lesions on peach associated with powdery mildew.

Stone Fruits

FUSICOCCUM CANKER

Fusicoccum canker, also called constriction canker, is caused by *Phomopsis amygdali* (Del.) Tuset and Portilla, formerly *Fusicoccum amygdali* Delacr. The disease is of minor economic importance on peach in the mid–Atlantic and south–eastern states and rare and of no economic importance on peach in Michigan.

Symptoms and Disease Cycle

Symptoms appear in early summer and become increasingly evident as more blighted shoots appear through late summer. Infected twigs and shoots wilt and die because of elongate, brown, sunken cankers, often with a zonate pattern, at their bases (*Photos 126, 127*). Constrictions formed at the bases of infected shoots and leaf symptoms produced well beyond the infection site result from a translocatable toxin, fusicoccin. Gumming is commonly associated with cankers, but it is not a good diagnostic characteristic because other canker pathogens also cause gumming.

Darkly pigmented pycnidia (flask–shaped, conidia–bearing fruiting bodies) are produced over the surface of the cankered area. The pycnidia exude conidia in white tendrils during wet weather. Conidia disseminated by rain infect through leaf scars in autumn and through buds, bud scale scars, stipule scars and fruit scars, or directly through young shoots during the growing season. The pathogen may also cause large, circular to irregular, zonate, brown spots in the leaves.

Photo 126. Death of peach shoots caused by Fusicoccum canker.

Control

Removing twigs with dieback and cankers is important to help control Fusicoccum canker. Fungicides, applied just before bud–break and in autumn, may be needed in problem orchards. Some fungicides used for brown rot and peach scab control during the growing season may help prevent infection. Cultivars vary in susceptibility, with Golden Jubilee, Redhaven, Rio–Oso–Gem and Redgold nectarine being more susceptible than Coronet, Harken and Sunhigh.

Photo 127. Close-up of constriction cankers at the bases of shoots blighted by *Phomopsis amygdali*. The cankers result from toxins produced by the pathogen.

Selected References

Guba, E. F. 1953. Large leaf spot and canker of peach caused by the fungus *Fusicoccum amygdali* Delacr. *Plant Dis. Rep.*, 37:560–564.

Roberts, J.W. 1940. The constriction disease of peach. *Phytopathology*, 30:963–968.

Tuset, J. J., and M. T. Portilla. 1989. Taxonomic status of *Fusicoccum amygdali* and *Phomopsis amygdalina*. *Can. J. Bot.*, 67:1275–1280.

GUMMOSIS

Gummosis was first reported in the United States on peach trees located near Fort Valley, Georgia, in the late 1960s and early 1970s, then later in neighboring states. It is caused by three species of *Botryosphaeria: B. dothidea, B. obtusa* and *B. rhodina.*

Photo 128. Extensive exudation of gum on the trunk of peach (gummosis) caused by *Botryosphaeria* spp.
(Courtesy P. L. Pusey, USDA-ARS, Washington Tree Fruit Research and Extension Center, Wenatchee)

Symptoms and Disease Cycle

Gummosis is characterized by numerous sunken necrotic lesions ¼ to ½ inch in diameter around lenticels and by excessive gum exudation (*Photo 128*). Removing the bark reveals shallow, round to oval, brown, gummy lesions ½ to 1 inch in diameter. On young branches, lenticels become swollen but gumming does not occur. Symptoms initially occur on the trunk between the ground and the scaffold limbs, usually during the second or third growing season. The disease later affects the scaffold limbs and twigs. Severe infection may kill twigs, reducing fruiting wood.

The *Botryosphaeria* spp. that cause this disease have many hosts. Conidia and ascospores produced in pycnidia and pseudothecia, respectively, in cankers on peach trees or other hosts are released during rain and are washed or blown onto peach branches. Infection through lenticels occurs primarily during the summer months. For identification of *B. dothidea* and *B. obtusa*, see apple white rot and black rot, respectively.

Control

Destroying and removing dead trees and prunings may be helpful in reducing inoculum. There is no chemical control for the disease.

Selected References

Brown, E. A., II, and K. O. Britton. 1986. *Botryosphaeria* diseases of apple and peach in the southeastern United States. *Plant Dis.,* 70:480–484.

Pusey, P. L. 1989. Availability and dispersal of ascospores and conidia of *Botryosphaeria* in peach orchards. *Phytopathology,* 79:635–639.

Pusey, P. L. 1989. Influence of water stress on susceptibility of nonwounded peach bark to *Botryosphaeria dothidea. Plant Dis.,* 73:1000–1003.

Reilly, C. C., and W. R. Okie. 1982. Distribution in the southeastern United States of peach fungal gummosis caused by *Botryosphaeria dothidea. Plant Dis.,* 66:158–161.

Weaver, D. J. 1974. A gummosis disease of peach trees caused by *Botryosphaeria dothidea. Phytopathology,* 64:1429–1432.

Stone Fruits

BACTERIAL CANKER

Two related bacteria, *Pseudomonas syringae* pv. *syringae* van Hall and *P. s.* pv. *morsprunorum* (Wormald) Young et al., can cause bacterial canker. Both pathogens affect sweet cherry, sour cherry, plums and prunes in Michigan and the neighboring province of Ontario, Canada. Disease outbreaks are sporadic and more frequent on sweet cherry than on sour cherry. *P. s. syringae* is found on peach in the southeastern United States.

Photo 129. Healthy Hardy Giant sweet cherry tree on left, tree with systemic bacterial canker infection on right. Note upright growth and discoloration of infected tree.

Symptoms

The disease attacks most parts of the tree (*Photo 129*). Cankers on trunks, limbs and branches exude gum during late spring and summer (*Photo 130*). Leaves on the terminal portions of cankered limbs and branches may wilt and die in summer or early autumn if girdled by a canker. Occasionally, large scaffold limbs are killed.

Leaf and fruit infections occur sporadically, but they can be of economic significance in years with prolonged wet, cold weather during or shortly after bloom. Leaf spots are dark brown, circular to angular, and sometimes surrounded with yellow halos (*Photo 131*). The spots may coalesce to form large patches of dead tissue, especially at margins of leaves, or the centers of the necrotic spots may drop out, resulting in tattered leaves. Infected leaves may abscise during midseason.

Lesions on green cherry fruit are brown with a margin of wet or water-soaked tissue (*Photo 132*). The affected tissues collapse, leav-

Photo 130. Gum exudation and upward extension of a canker on sweet cherry caused by *Pseudomonas syringae* pv. *morsprunorum*.

Photo 131. Sour cherry with necrotic spots and yellowing of leaves due to bacterial canker.

ing deep, black depressions in the flesh, with margins becoming yellow to red as lesions and fruit age. On fruit stems, lesions are elliptical and brown with water-soaked margins.

Infected leaf and flower buds may fail to open in spring, resulting in a condition referred to as "dead bud." Small cankers often develop at the bases of these buds. Other infected buds open in spring but collapse in early summer, leaving wilted leaves

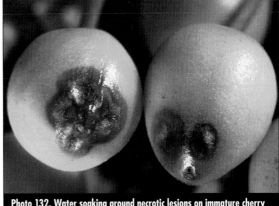

Photo 132. Water soaking around necrotic lesions on immature cherry fruit caused by bacterial canker.

and dried-up fruit. If blossom infection occurs, whole blossom clusters collapse as infection spreads into the fruit-bearing spurs (*Photo 133*). Blossom blight and spur blast are most likely in years when leaf and fruit infections are common.

Disease Cycle

The bacteria can survive from one season to the next in bark tissue at canker margins, in apparently healthy buds and systemically in the vascular system. Bacteria multiply within these overwintering sites in the spring and are disseminated by rain to blossoms and to young leaves. Bacteria of both pathovars can live in an epiphytic phase on the surface of symptomless blossoms and leaves from bloom through leaf fall in autumn (*Photo 134*). After leaves abscise in autumn, the bacteria may enter the tree through fresh leaf scars.

Outbreaks of bacterial canker are often associated with prolonged periods of cold, frosty, wet weather late in the spring or with severe storms that injure the emerging blossom and leaf tissues. Freezing can predispose the tissue to infection, but infection depends on the presence of wet weather during the thawing process. Free water on leaf surfaces and high relative humidity are required for at least 24 hours before significant leaf infection can occur following violent storms. Symptoms appear about 5 days later at temperatures between 70 and 80 degrees F.

Control

The disease is troublesome on some sweet cherry cultivars but not others. Schmidt and Windsor are susceptible and often severely damaged. Hardy Giant is very susceptible—this cultivar should be avoided.

Copper-containing compounds may be of limited value for the control of bacterial canker because strains of *P. s. syringae* resistant to copper are common in orchards with a history of copper usage. Also, copper injures most stone fruit crops. Even on the more tolerant crop species, it becomes more injurious as applications are repeated.

Photo 133. Spur dieback and necrosis of the midvein of Stanley plum caused by bacterial canker.

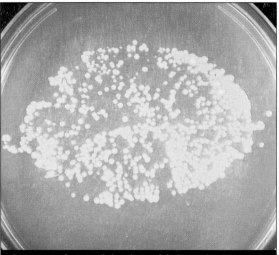

Photo 134. Leaf print made by colonies of the bacterial canker organism on a culture medium demonstrates the presence of bacteria on the surface of an apparently healthy leaf.

Selected References

Allen, W. R., and V. A. Dirks. 1978. Bacterial canker of sweet cherry in the Niagara Peninsula of Ontario: *Pseudomonas* species involved and cultivar susceptibilities. *Can. J. Plant Sci.,* 58:363–369.

Dhanvantari, B. N. 1969. Occurrence of bacterial canker of sweet cherry and plum in Ontario. *Can. Plant Dis. Surv.,* 49:5–7.

Jones, A. L. 1971. Bacterial canker of sweet cherry in Michigan. *Plant Dis. Rep.,* 55:961–965.

Latorre, B. A., and A. L. Jones. 1979. *Pseudomonas morsprunorum,* the cause of bacterial canker of sour cherry in Michigan and its epiphytic association with *P. syringae. Phytopathology,* 69:335–339.

Paterson, J. M., and A. L. Jones. 1991. Detection of *Pseudomonas syringae* pv. *morsprunorum* on cherries in Michigan with a DNA hybridization probe. *Plant Dis.,* 75:893–896.

Sundin, G. W., A. L. Jones and D. W. Fulbright. 1989. Copper resistance in *Pseudomonas syringae* pv. *syringae* from cherry orchards and its associated transfer *in vitro* and *in planta* with a plasmid. *Phytopathology,* 79:861–865.

Sundin, G. W., A. L. Jones and B. D. Olson. 1988. Overwintering and population dynamics of *Pseudomonas syringae* pv. *syringae* and *P. s.* pv. *morsprunorum* on sweet and sour cherry trees. *Can. J. Plant Pathol.,* 10:281–288.

Stone Fruits

SILVER LEAF

Silver leaf, caused by the fungus *Chondrostereum purpureum* (Pers. ex.:Fr.) Pouzar, occurs occasionally on isolated trees in sour cherry and apple orchards throughout the northeastern United States and Canada. It is of negligible economic importance.

Symptoms and Disease Cycle

Leaves, usually on one or two branches per tree, exhibit a dull metallic luster (*Photo 135*). Affected leaves may also exhibit necrotic areas. These "silver leaves" are due to a separation of the upper epidermis from the palisade layer. Because of this separation, the upper epidermis can be peeled back easily.

The causal fungus is located in the wood of the trunk or branches beneath affected leaves. The fungus decays the wood and produces a toxin that induces the silvery condition when translocated to leaves.

The vigor of infected trees declines for several years before they die. Eventually, flat, grayish brown fruiting bodies of *C. purpureum* may be observed on the bark of dead branches and trees. Basidiospores, produced in the fruiting bodies from autumn to early summer the following year, germinate and infect primarily through fresh wounds. Then, the fungus extends upward and downward from infected wounds.

Photo 135. Silver leaf symptoms on apple leaves. Symptoms on cherries are similar.

Control

Maintaining trees in good vigor, pruning off broken and dead limbs, and training and pruning trees annually–thereby avoiding the need to make major pruning cuts later–all help to prevent infection and reduce disease spread.

Selected References

Fujita, K. 1990. Silver Leaf. Pages 42–43 in: *Compendium of Apple and Pear Diseases.* A. L. Jones and H. S. Aldwinckle, eds. St. Paul, Minn.: American Phytopathological Society.

Stone Fruits

PEACH PERENNIAL CANKER

Perennial canker of peach, also called Leucostoma canker, Cytospora canker and Valsa canker, is caused by two related fungi, *Leucostoma cincta* (Fr. ex.:Fr.) Höhn and *L. persoonii* Höhn. The disease is particularly important and common in peach orchards in the northeastern states and can be found throughout the eastern United States. It is sporadic and of lesser importance on nectarines, and minor on apricot, prune, plum and sweet cherry.

Symptoms

Cankers on branches, scaffold limbs, branch crotches and trunk are oval to elliptical in outline and often are surrounded by a roll of callus at the margins (*Photo 136*). Characteristic white, circular stromata of the fungus are visible when the bark is removed over the cankered area (*Photo 137*). Infected 1–year–old shoots, partic- ularly weak shoots on scaffold limbs, exhibit dieback. Water– soaked tissue and gumming may be observed on branches and limbs at the bases of infected shoots. Later, cankers may develop, each with a dead shoot in the center.

Cankers gradually enlarge each year until the limb or trunk is completely girdled. Active cankers often have gum associated with them, but "gummosis" by itself is not diagnostic. Leaves on limbs girdled by cankers collapse and die. Old cankers are rough and charcoal–black.

Disease Cycle

The fungi overwinter in cankers or on deadwood. Pycnidia are located in diseased tissue under the bark. Conidia, produced in tendrils extruded from the pycnidia, are disseminated by splashing and wind– driven rain. Infection is through damaged or injured bark. Cold injury is the most important factor predisposing trees to canker; pruning wounds, mechanical damage, insect punctures and leaf scars are other entry points.

Moisture is required for spore germination. The rate of canker development after infection depends on temperature and the species of fungus involved. *Leucostoma cincta* is favored by lower temperatures than *L. persoonii*. Optimum temperatures for growth of *L. cincta* and *L. persoonii* are approximately 68 and 86 degrees F, respectively. When temperatures are not favorable for fungal activity, callus formation occurs. Canker activity resumes when temperatures again favor the fungus.

Control

No single measure· is adequate to control perennial canker. Control measures involve cultural practices that are aimed at reducing the susceptibility of peach trees to canker and inoculum in and around the orchard.

Cultural practices that help to reduce the incidence and severity

Photo 136. Perennial canker of peach.

Photo 137. Peach canker with bark removed to expose black structures with white margins containing the pycnidial stromata of *Leucostoma* spp.

of perennial canker include planting new orchards on well drained sites and using management practices that promote winter hardiness. Ways to promote early cessation of growth without excessive loss of tree vigor include applying fertilizer early, planting a cover crop by July 1 in cultivated orchards and mowing thereafter as needed. Applying white latex paint to the southwest side of trunks and lower scaffold branches also helps to prevent cold injury and avoid the disease.

Inoculum can be reduced by removing and destroying badly cankered limbs, branches or trees and by planting new peach orchards away from older orchards heavily infected with peach canker. Sanitation is a must during the early life of the orchard.

Delay orchard pruning until growth starts in spring–late spring pruning promotes quick healing. Moderate to severe pruning in November, or earlier, can severely weaken or kill trees. Try to avoid mechanical and insect injury and do not leave long pruning stubs. Avoid weak–angled crotches when shaping trees.

Apply fungicide sprays after pruning and before rain.

Selected References

Adams, G., S. A. Hammar and T. J. Proffer. 1990. Vegetative compatibility in *Leucostoma persoonii*. *Phytopathology*, 80:287–291.

Biggs, A. R. 1989. Integrated approach to controlling Leucostoma canker of peach in Ontario. *Plant Dis.*, 73:869–874.

Dhanvantari, B. N. 1982. Relative importance of *Leucostoma cincta* and *L. persoonii* in perennial canker of peach in south western Ontario. *Can. J. Plant Pathol.*, 4:221–225.

Surve-Iyer, R. S., G. C. Adams, A. F. Iezzoni and A. L. Jones. 1995. Isozyme detection and variation in *Leucostoma* species from *Prunus* and *Malus*. *Mycologia*, 87:471–482.

Tekauz, A., and Z. A. Patrick. 1974. The role of twig infections on the incidence of perennial canker of peach. *Phytopathology*, 64:683–688.

ANTHRACNOSE

Anthracnose, also called ripe rot, is caused by the fungus *Colletotrichum gloeosporioides* (Penz.) Penz. and Sacc. and *C. acutatum* J. H. Simmonds. It is rare and of no economic importance on peach in southwestern Michigan and absent elsewhere in Michigan. It is a minor disease of peach in areas of the eastern United States where bitter rot is a significant problem on apples.

Symptoms and Disease Cycle

Anthracnose begins as small, brown spots on ripening fruit that slowly enlarge to 1 inch or more in diameter. Early symptoms of anthracnose may be confused with brown rot; however, brown rot lesions enlarge much faster than those of anthracnose.

Older lesions are circular, brown, firm, slightly sunken and often covered with concentric rings of saucer–shaped fungal fruiting bodies (acervuli) bearing salmon–colored conidial masses (*Photo 138*). For a more complete description of the pathogen, see bitter rot in the section on pome fruit.

Anthracnose is most likely to occur in seasons with above normal rainfall in summer prior to harvest. Ascospores and conidia, often from other hosts, initiate the disease during wetting periods suitable for dissemination and germination of spores. A few primary infections to fruit scattered throughout the orchard provide the inoculum for secondary spread of the disease.

Control

Fungicides for brown rot and peach scab that are also active against anthracnose should be used in the cover sprays where anthracnose is a potential problem.

Selected References

Bernstein, B., E. I. Zehr, R. A. Dean and E. Shabi. 1995. Characteristics of *Colletotrichum* from peach, apple, pecan, and other hosts. *Plant Dis.*, 79:478–482.

Ramsey, G. B., M. A. Smith and B. C. Heiberg. 1951. Anthracnose of peaches. *Phytopathology*, 41:447–455.

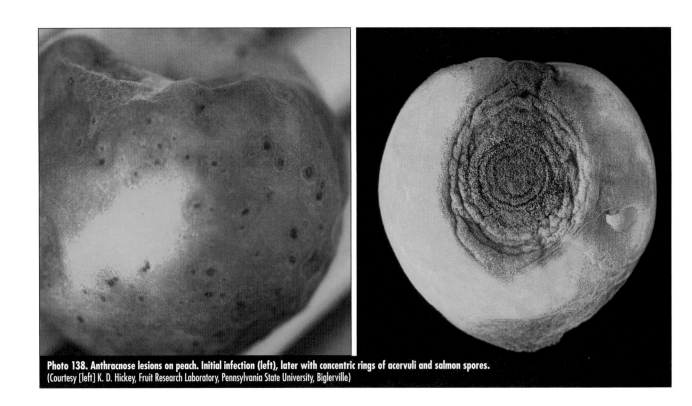

Photo 138. Anthracnose lesions on peach. Initial infection (left), later with concentric rings of acervuli and salmon spores.
(Courtesy [left] K. D. Hickey, Fruit Research Laboratory, Pennsylvania State University, Biglerville)

Stone Fruits

RHIZOPUS ROT

Rhizopus rot, caused by the fungus *Rhizopus stolonifer* (Ehrenb.) Vuill., occurs primarily on ripe peach, nectarine, sweet cherry and plum fruit. The disease is commonly observed on fruit in storage, transit or marketplace. It is usually of minor importance in the field but sometimes causes important postharvest losses.

Symptoms and Disease Cycle

Rhizopus causes a brown to black decay or rot of injured or overmature fruit. Symptoms are rarely observed on the tree, but Rhizopus rot may decay fruit on the orchard floor.

The skin covering lesions caused by *R stolonifer* readily slips from the flesh underneath (*Photo 139*). Affected fruit are soft, watery and often covered with a whisker–like growth. Black, spherical structures (sporangia), each containing thousands of small sporangiospores, are produced at the tips of the whisker–like growth. The outer wall of the sporangia ruptures easily, freeing the sporangiospores, which are disseminated to other fruit in air currents.

Once the pathogen becomes established on fruit, it can spread onto sound fruit nearby, resulting in clusters of infected fruit. The fungus invades quickly, producing tufts of whisker–like fungal growth over the surface of the clusters.

Photo 139. Rhizopus rot on peach.
(Courtesy K. D. Hickey, Fruit Research Laboratory, Pennsylvania State University, Biglerville)

Control

Rhizopus rot is effectively controlled by storing fruit at or below 39 degrees F because the fungus does not grow below 40 degrees F. Careful handling of the fruit to avoid wounds, clean storage containers and warehouse facilities, hydrocooling with clean water, use of preharvest fungicide sprays, and postharvest fungicide dips, sprays or impregnated wraps help to prevent the disease. Only one or two fungicides are efficacious against the pathogen, and these may not be approved for use on all fruit crops.

Selected References

Daines, R. H. 1965. 2,6-dichloro-4-nitroaniline used in orchard sprays, and dump tank, the wet brusher and the hydrocooler for control of Rhizopus rot of harvested peaches. *Plant Dis. Rep.*, 49:300–304.

GILBERTELLA ROT

Gilbertella rot of peach, caused by the fungus *Gilbertella persicaria* (E. D. Eddy) Hasseltine, was first reported in the southeastern United States in 1992. Signs of the pathogen differ from those of *Rhizopus*, but symptoms are virtually identical and the two diseases are easily confused.

Symptoms of Gilbertella rot usually are not visible on fruit on the tree before harvest but develop on fruit that fall to the orchard floor or on fruit during storage. The rot begins as small, tan lesions that rapidly enlarge to affect the entire fruit. Affected fruit are soft and watery and the skin covering lesions easily slips from the flesh beneath it (Photo 140). Unlike *Rhizopus*, the mycelium does not grow out from the fruit in whisker–like growth but remains lying closely against the skin. As the rot progresses, the surface of the fruit becomes black with numerous sporangiophores produced on its surface.

Control

Controls for Gilbertella rot are similar to those for Rhizopus rot. Except for chlorination, however, preharvest sprays and postharvest treatments are not effective. Careful attention to postharvest sanitation and chlorination of water in the packing house are necessary for control of *G. persicaria*.

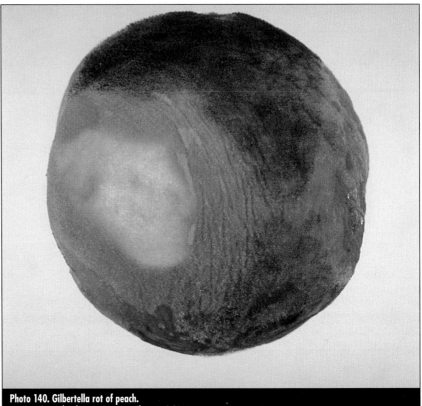

Photo 140. Gilbertella rot of peach.
(Courtesy E. I. Zehr, Clemson University, Clemson, S.C.)

Selected References

Ginting, C., and Zehr, E. I. 1992. Gilbertella rot of peaches caused by *G. persicaria* in South Carolina. *Plant Dis.*, 76:753.

Stone Fruits

ALTERNARIA FRUIT ROT

Alternaria fruit rot, caused by a species of the fungus *Alternaria*, is a minor problem on sweet and sour cherries throughout the northeastern United States. The disease is most severe on overripe fruit or where rain-induced cracking or various physical injuries expose the flesh to infection.

Lesions are circular to oblong and slightly sunken, later becoming firm, flattened and wrinkled, and often dark green to black because of abundant sporulation by the pathogen (*Photos 141, 142*). Lesions may cover one-third to one-half of the fruit. The conidia are dark and typically produced in chains.

In cold storage, rot often starts in the cavity left when the stem is detached from the fruit, in cracks of the skin and flesh, or in bruised areas. The fungus produces fluffy, gray to white strands of mold over the surface of the lesions.

No effective control measures are available except to harvest fruit before they are injured or overripe.

Photo 141. Alternaria rot on dark sweet cherry fruit. Note flattened and wrinkled nature of the lesions.

Photo 142. Decay of Gold sweet cherry fruit due to Alternaria rot.

Peach tree short life is a serious disease affecting peaches in the southeastern United States. It is a disease complex involving several pathogens, environmental factors and cultural practices.

Symptoms

The disease is characterized by the sudden death of trees above the soil line after bloom or during early growth. In some instances, trees may fail to bloom or leaf out, or symptoms may be delayed until early summer. Affected trees often appear healthy the previous fall. Droplets of yellow or orange ooze may be visible on the bark of the trunk or main scaffold limbs. The inner bark and cambium are discolored and may have a sour sap odor.

Cause and Contributing Factors

Peach tree short life is caused by freeze injury or *Pseudomonas syringae* pv. *syringae* (the cause of bacterial canker) or a combination of the two. Affected tissues are often colonized by a species of the fungus *Cytospora* (the cause of perennial canker). Rootstocks, time of pruning, soil fumigation, nematodes and characteristics of the orchard site are involved in disease incidence. Trees propagated on root-knot nematode-resistant rootstocks (i.e., Nemaguard) are more susceptible than those planted on Lovell or Halford. A new peach rootstock, Guardian (formerly BY520-9), looks promising for preventing peach tree short life. In trials in Georgia and South Carolina, trees budded on Guardian survived better than those on Lovell or Nemaguard seedlings on sites where root-knot nematode and peach tree short life have been problems. Trees pruned in the autumn are much more susceptible than those pruned in the late winter or early spring.

The disease is less severe where preplant and postplant fumigation are used. High populations of ring nematodes are associated with peach tree short life. The disease is usually more severe where trees are planted in old orchard sites.

Disease Identification

Peach tree short life is usually diagnosed by the characteristic symptoms associated with it: sudden death and collapse of the tree in the spring. Freeze injury is characterized by a brown discoloration of the cambial layer and, in severe instances, cracking and separation of the bark from the wood.

Pseudomonas syringae pv. *syringae* infection is difficult to distinguish from freeze injury. Infected bark is red-brown and cankers have a definite margin. When injured tissues are invaded by *Cytospora* spp., fruiting structures (pycnidia) are usually visible just beneath the bark. Under moist conditions, conidia produced in thread-like strands extrude from the pycnidia.

Control

A 10-point program has been developed for control of peach tree short life. The three most important elements of the program are:

- Prune as late in spring as possible.
- Plant trees propagated on Guardian, Lovell or Halford rootstock.
- Use preplant soil fumigation.

Other elements of the program are postplant soil treatment, liming for proper pH, subsoiling before planting, maintaining a good fertilization program, using herbicides or shallow cultivation, and removing all dead or dying trees promptly.

Selected References

Ritchie, D. F., and C. N. Clayton. 1981. Peach tree short life: A complex of interacting factors. *Plant Dis.*, 65:462–469.

Stone Fruits

CROWN GALL

Crown gall, caused by the bacterium *Agrobacterium tumefaciens* (E. F. Smith and Townsend) Conn, affects a wide range of herbaceous and woody plant species, including stone and pome fruit trees. The greatest losses occur in nurseries; losses in orchards are sporadic and minor, provided that nursery trees are symptomless when planted.

Symptoms

Crown gall is characterized by the formation of tumors or galls on the roots and crown and, under special circumstances, on aboveground portions of plants. Galls begin as small, smooth growths that enlarge to become dark, hard, woody tumors with gnarled, irregular surfaces. Tumors range from ¼ inch to more than 4 inches in diameter (*Photo 143*). Galls are typically globular but may be elongated or otherwise irregular. Several galls may occur on the same root or crown. Old galls may be covered with secondary fungi and riddled by insects. Orchard trees with one or more large galls on the crown are often stunted.

Disease Cycle

The pathogen is widely distributed in many nursery and some orchard soils, particularly soils used to grow susceptible plants within the past 5 years. Another source of the pathogen is symptomless but infected nursery stock. The bacteria enter the roots and crown primarily through wounds produced in caring for and handling the nursery stock.

Symptoms may develop in a few weeks at moderate tempera-tures, or the bacterium may remain latent for 2 to 3 years before symptoms are produced. If crown gall occurs in the nursery, symptoms are usually well developed on finished trees at the time of digging. The bacterium carries a large, circular piece of DNA known as the tumor–inducing (Ti) plasmid. When bacteria infect plants, a portion of the Ti plasmid is inserted into the chromosome of a plant cell. Expression of this genetic material in the host results in overproduction of plant hormones, which in turn causes gall formation.

Control

Planting disease–free nursery stock is extremely important to avoid problems with crown gall in the orchard. All nursery stock should be inspected and any infected trees discarded or returned to the nursery.

Many materials and methods have been tested to control crown gall over the past 75 years, but the most successful has been based on strain 84 of the antagonistic bacterium *A. radiobacter* (Beijerinck and van Delden) Conn. When used on seeds, roots and stems of propagation materials, this biological control has proved very effective in preventing crown gall on all tree fruit species except apple (*Photo 144*).

Selected References

Dhanvantari, B. N. 1976. Biological control of crown gall of peach in south-western Ontario. *Plant Dis. Rep.*, 60:549–551.

Kerr, A. 1980. Biological control of crown gall through production of agrocin 84. *Plant Dis.*, 64:25–30.

Moore, L. W., and G. Warren. 1979. *Agrobacterium radiobacter* strain 84 and biological control of crown gall. *Ann. Rev. Phytopathol.*, 17:163–179.

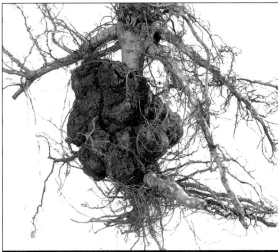

Photo 143. Crown gall of Mazzard F12/1 rootstock.

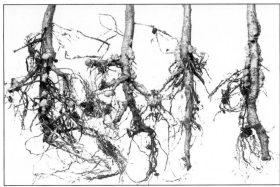

Photo 144. Biological control of crown gall on *Prunus* seedlings with *Agrobacterium radiobacter* strain 84. Treated seedlings (bottom) were dipped into a solution of strain 84, then after 24 hours both sets of seedlings were inoculated with *Agrobacterium tumefaciens*, the crown gall pathogen.

Stone Fruits

VERTICILLIUM WILT

Verticillium wilt, caused by the fungus *Verticillium dahliae* Kleb., is an occasional problem on sour cherry and a rare problem on sweet cherry, peach, prune and plum trees in the northeastern United States. The fungus that causes this disease is a common pathogen of strawberry, potato, tomato, several other vegetable crops and many species of weeds.

Symptoms and Disease Cycle

The first indication of Verticillium wilt is an unexplained flagging of leaves on one or more branches in midsummer. Wilting of leaves is followed by yellowing and curling of the leaves and, ultimately, defoliation *(Photo 145)*. Symptoms usually develop first on the lower parts of shoots and progress upward, leaving a few green leaves at the tip. Some branches are killed, but diseased trees, sometimes stunted and unproductive for a few years, eventually recover. Brownish gray streaks may be observed in the sapwood of branches with symptoms *(Photo 146)*.

The soilborne fungus can survive as microsclerotia for several years in the absence of a susceptible host. Soil populations of the pathogen can build up to very high levels on susceptible weeds such as lamb's-quarter, pigweed, nightshade and ground cherry, and on such crops as strawberries, raspberries, potatoes, tomatoes, peppers and eggplants. The fungus invades through the roots of trees and then develops in the vascular systems. In Michigan, the disease has been found primarily in orchards planted on old potato fields or where contaminated soil from potato fields was carried into orchards by an overflow of irrigation water.

Control

Control measures consist largely of avoiding old potato, tomato and strawberry fields as planting sites and not interplanting orchards with these crops. Excess irrigation water from adjacent fields with a susceptible crop such as potatoes should not be allowed to flow into orchard plantings.

Selected References

Parker, K. G. 1959. *Verticillium hadromycosis* of deciduous tree fruits. *Plant Dis. Rep. Supplement*, 255:39–61.

Photo 145. Defoliation and wilting of lateral branches of Montmorency sour cherry caused by Verticillium wilt.

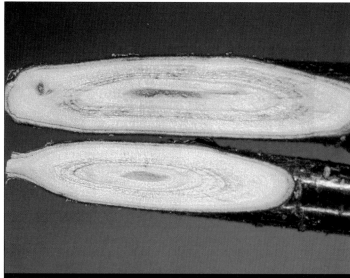

Photo 146. Grayish streaks in the sapwood of Montmorency sour cherry with Verticillium wilt.

Stone Fruits

ARMILLARIA AND CLITOCYBE ROOT ROTS

Armillaria root rot is a soilborne disease that occurs throughout the stone fruit–growing regions of the eastern United States. In Michigan, it is increasing as a problem on Montmorency sour cherry. Clitocybe root rot is a closely related soilborne disease that occurs mainly in peach–growing regions of the southeastern United States. Both diseases are also destructive on many ornamental and forest trees.

Several species of Armillaria (*A. mellea sensu stricto* [Vahl ex Fr.] Kummer, *A. ostoyae* [Romagn.] Herink., North American biological species III, and *A. bulbosa* [Barla] Kile and Watling) cause Armillaria root rot on sour cherry, sweet cherry and peach in Michigan. These fungi are common soil inhabitants in woodlands throughout the eastern United States. *Armillaria tabescens* (Scop.) Dennis et al., previously named *Clitocybe tabescens* (Scop.) Bres., causes Clitocybe root rot.

Symptoms

Armillaria root rot is most severe on trees located on sandy, well drained soils. Affected trees, often with reduced terminal shoot growth, exhibit reddish to purplish foliage much earlier in the autumn than healthy trees. Diseased trees are prone to collapse suddenly in midsummer, and the leaves fail to abscise and remain attached into early winter (*Photo 147*). Trees usually die in

a circular pattern from foci consisting initially of one or two infected trees.

Proper diagnosis requires removing soil from around the bases of declining trees or pulling trees from the soil. Outer bark tissues removed from the crown or roots are often soft and necrotic. Thick, white fungal growth is often seen in the necrotic tissue, and a fan–shaped fungal mat is often present between the necrotic inner bark and the wood. *Armillaria* is distinguished from other fungi by the presence of dark brown to black, thread–like structures called rhizomorphs (*Photo 148*). These structures are about the size of a shoestring. Clusters of mushrooms may arise at the bases of dead trees in late August or September. These mushrooms are honey-colored with a ring or annulus on the stem or stipe just beneath the gills (*Photo 149*).

Unless mushrooms or rhizomorphs are present, Clitocybe root rot is not easily distinguished from Armillaria root rot. The presence of perforations in the white mycelial mat of *Clitocybe* is an aid in separating the two diseases. Rhizomorphs are not formed. Rather, black, hardened mycelial extrusions often develop through cracks in the bark. In autumn during rainy periods, clusters of yellow to yellow-brown mushrooms may arise at the bases of infected

Photo 147. Sudden collapse of a Montmorency sour cherry tree in midsummer caused by Armillaria root rot. Leaves remain attached into winter.

Photo 148. Armillaria root rot. White fungal mat beneath the bark (left) and necrotic tissue with white fungal growth beneath the bark and rhizomorphs in the surrounding soil (right).

Photo 149. Mushrooms of *Armillaria ostoyae* around the base of a cherry trunk.

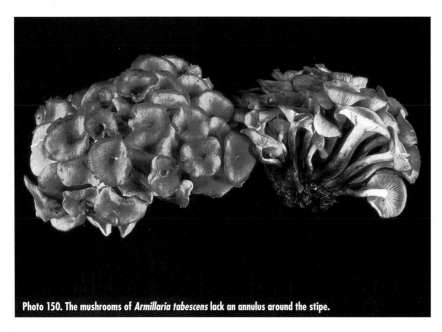
Photo 150. The mushrooms of *Armillaria tabescens* lack an annulus around the stipe.

Selected References

Anderson, J. B. 1983. Bifactorial heterothallism and vegetative diploidy in *Clitocybe tabescens*. *Mycologia*, 74:911–916.

Anderson, J. B. 1986. Biological species of *Armillaria* in North America: Redesignation of groups IV and VII and enumeration of voucher strains for other groups. *Mycologia*, 78:837–839.

Guillaumin, J. J., J. Pierson and C. Grassely. 1991. The susceptibility to *Armillaria mellea* of different *Prunus* species used as stone fruit rootstocks. *Scientia Horticulturae*, 46:43–54.

Proffer, T. J., A. L. Jones and G. R. Ehret. 1987. Biological species of *Armillaria* isolated from sour cherry orchards in Michigan. *Phytopathology*, 77:941–943.

Proffer, T. J., A. L. Jones and R. L. Perry. 1988. Testing of cherry rootstocks for resistance to infection by species of *Armillaria*. *Plant Dis.*, 72:488–490.

Rhoads, A. S. 1954. Clitocybe root rot found widespread and destructive in Georgia and South Carolina peach orchards. *Plant Dis. Rep.*, 38:42–46.

trees (*Photo 150*). These mushrooms, unlike those of other species of *Armillaria*, do not have an annulus around the stipe.

Disease Cycle

The fungi that cause Armillaria and Clitocybe root rots may exist in an orchard site on roots from previously infected orchard and forest trees. Less frequently, they may be introduced into orchards in contaminated soil or equipment.

Mycelial strands in root debris and rhizomorphs in soil or associated with root debris survive in the soil for many years. Healthy roots are infected when they come in contact with fungal extrusions or with mycelium in diseased roots. Rhizomorphs also function in spreading the fungus from tree to tree. It has been estimated that a colony of *Armillaria* has existed in a Michigan forest for at least 1,500 years.

Control

Newly cleared woodland or orchard sites with a history of Armillaria or Clitocybe root rots should not be planted to stone fruit crops. If replanting of infested tree sites is attempted, the soil should be clean cultivated and as much root debris as possible removed. Fumigation has not been very effective in controlling the disease because rhizomorphs and infected root debris deep in the soil are not easily reached and permeated by chemical vapors.

Rootstocks of the common stone fruit crops grown in the eastern United States are highly susceptible, apple rootstocks are moderately susceptible and pear rootstocks are considered tolerant to *Armillaria*. Sweet cherry cultivars on Mazzard rootstocks are more resistant to *Armillaria* than sour cherry cultivars on Mazzard rootstocks, and all cultivars of sweet and sour cherry on Mahaleb rootstock are highly susceptible to *Armillaria*. Some rootstocks resistant to *Armillaria* in one test may be susceptible in another test because of differences in the species of *Armillaria* involved.

Stone Fruits

PHYTOPHTHORA ROOT AND CROWN ROT

Phytophthora root and crown rot was identified in Michigan and in other areas of the eastern United States in the 1980s. Tree losses were severe at problem sites, but only a few orchards were affected in each area.

Several species of *Phytophthora* (*P. megasperma* Drechsler, *P. cryptogea* Pethyb. and Laff., *P. cambivora* [Petri] Buisman, *P. syringae* [Kleb.] Kleb., *P. cactorum* [Lebert and Cohn] Schroet., and an unidentified *Phytophthora* sp.) were identified as causal agents of the disease in cherry trees. *Pythium* species commonly associated with cherry roots were not found to be primary causal agents in the decline and death of cherry trees. In New York and Ohio, *P. megasperma*, *P. cryptogea* and *P. cactorum* were identified as the causal agents on peach trees.

Symptoms

Trees affected with Phytophthora root and crown rot often are concentrated in areas of the orchard where water tends to accumulate on the surface of the soil, where a hardpan hinders drainage in the lower root zone or where heavy soils drain slowly (*Photo 151*). The disease is most likely to appear when the trees come into production, generally 4 to 6 years after planting.

Affected trees exhibit poor terminal growth, sparse and chlorotic foliage, and progressive decline over several seasons. Some trees exhibit early reddish discoloration of leaves in late

Photo 151. Progressive decline and eventual death of Montmorency sour cherry trees from Phytophthora root and crown rot.

August or early September. A few trees collapse and die soon after budbreak.

Proper diagnosis requires removing soil from the bases of declining trees or pulling the affected trees. After the outer bark is removed with a knife, necrotic tissue should be observed on the primary or secondary roots of declining trees; necrosis may extend to the soil line but seldom above it (*Photo 152*). Occasionally, infections are associated only with necrotic lesions on the crown.

Disease Cycle

Phytophthora is a soilborne fungus that thrives when the soil is saturated with water. It is an ubiquitous soil inhabitant or may be introduced into an orchard on infested nursery stock. It persists in orchards for several years, pri-

Photo 152. Decay of Mahaleb rootstock below ground caused by Phytophthora root and crown rot.

marily as oospores in infected host tissue, debris or soil. The oospores germinate when the soil moisture levels are high or saturated, provided temperatures are favorable. Sporangia produced by germinating oospores quickly produce zoospores. The zoospores swim or are moved passively by moving water and, once close to trees, actively infect the roots and crown.

Control

The most effective control strategy for preventing Phytophthora root and collar rot is to eliminate wet soil conditions through cultural practices such as site selection, use of drainage systems and raised planting beds. Mazzard cherry rootstocks are more resistant than Mahaleb cherry rootstocks to some species of *Phytophthora*. The effectiveness of resistant rootstocks or of fungicides is limited, however, because several species of *Phytophthora* are often present in the same orchard. Rootstocks and fungicides control some species of *Phytophthora* but not others.

Selected References

Bielenin, A., and A. L. Jones. 1988. Prevalence and pathogenicity of *Phytophthora* spp. from sour cherry trees in Michigan. *Plant Dis.*, 72:473–476.

Bielenin, A., S. N. Jeffers, W. F. Wilcox and A. L. Jones. 1988. Separation by protein electrophoresis of six species of *Phytophthora* associated with deciduous fruit crops. *Phytopathology*, 78:1402–1408.

Mircetich, S. M., and M. E. Matheron. 1976. Phytophthora root and crown rot of cherry trees. *Phytopathology*, 66:549–558.

Smither, M. L., and A. L. Jones. 1989. *Pythium* species associated with sour cherry and the effect of *P. irregulare* on the growth of Mahaleb cherry. *Can. J. Plant Pathol.*, 11:1–8.

Wilcox, W. F., and M. A. Ellis. 1989. Phytophthora root and crown rot of peach in the eastern Great Lakes region. *Plant Dis.*, 73:794–798.

Wilcox, W. F., and S. M. Mircetich. 1985. Pathogenicity and relative virulence of seven *Phytophthora* spp. on Mahaleb and Mazzard cherry. *Phytopathology*, 75:221–226.

Stone Fruits

X-DISEASE

X–disease is a serious disease of peach, nectarine, sweet cherry and sour cherry in the Great Lakes states and the province of Ontario, Canada. It is an occasional problem on peach and nectarine in the mid–Atlantic states, and it can be found on wild chokecherry in the Great Smoky Mountains as far south as northern Georgia.

X–disease is caused by a phytoplasma, a group of small, parasitic organisms formerly known as mycoplasma–like organisms, or MLOs. Phytoplasmas live in phloem cells of plants as spherical to ellipsoid–shaped bodies that are slightly smaller than bacteria (*Photo 153*). Besides X–disease, other phytoplasmas have been associated with peach yellows, little peach and peach rosette.

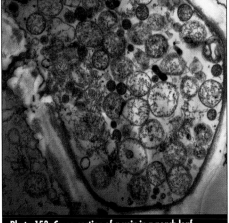

Photo 153. Cross-section of a vein in a peach leaf infected with X-disease. The spherical bodies are phytoplasmas and are found in a phloem cell.

Photo 154. Peach leaves with rolling, red blotch and tattering due to X-disease. Note defoliation starting at the base of the shoot.

Photo 155. Peach leaves yellowed and rolled from X-disease in late August.

Photo 156. Symptoms of X-disease randomly distributed throughout tree.

Symptoms

New infections appear on one or two limbs; 2 to 3 years later, symptoms may occur on several limbs. In peach, leaves on isolated limbs curl inward after about 2 months of growth and develop irregular yellow to reddish purple spots (*Photos 154–156*). The centers of the spots soon drop out, resulting in tattered leaves. Leaves on affected limbs fall prematurely, starting at the bases of the shoots. Only tufts of leaves remain at the tips of infected shoots. Fruit set may appear normal at first, but fruit on infected limbs will drop prematurely.

Cherries on Mahaleb root-stock die suddenly in mid–summer for no apparent reason. Trees on Mazzard rootstock decline slowly. Scattered fruit on trees on Mazzard rootstock are smaller than normal, are green or pink at harvest because they ripen slowly and have a bitter taste (*Photo 157*). Enlarged stip-ules may be associated with leaves of some sweet cherry cultivars with X–disease on Mazzard rootstock (*Photo 158, page 88*).

Photo 157. Green to half-ripe cherries interspersed with normal fruit on a sour cherry tree on Mazzard rootstock infected with X-disease.

Spread of X–Disease

The X-disease phytoplasma is transmitted by several species of leafhoppers (*Photo 159*). These leafhoppers are usually not considered pests of peach and cherry. They feed during the day on plants in the orchard groundcover and at dusk migrate to cherry and peach trees, where they remain until dawn. The main damage they cause is the incidental transmission of X–disease.

Leafhoppers acquire phytoplasma from the leaves of X-disease–infected chokecherry, sweet cherry or sour cherry. Two to 3 weeks later, they can transmit the agent to healthy leaves while feeding. Tree-to-tree spread from cherry trees with X–disease to adjacent peach orchards can be particularly important. Although the possibility of spread from peach to peach has been investigated many times, it appears to be of minor importance.

Control

Eradication of chokecherry near stone fruit orchards helps to control X–disease. Chokecherry bushes are commonly found in hedgerows, along property lines, in open woods, and in overgrown meadows and abandoned fields. Infected cherry trees, particularly those on Mazzard rootstock, should also be removed.

Brush killers offer the cheapest and most effective way to kill chokecherry bushes with both summer and autumn spray treatments. Other removal methods include bulldozing, deep plowing, burning or pulling out individual bushes.

In each growing season following removal, check the treated area carefully for chokecherry sprouts. Sprouts or new chokecherry seedlings should be treated with herbicide sprays or pulled out.

Maintaining a vigorous insect control program from June through harvest with insecticides effective against leafhoppers may also help to reduce the spread of X–disease.

Temporary symptom remission can be obtained with postharvest injections of oxytetracycline into the trunk. Sprays of oxytetracycline have not been as effective as injection treatments.

In nurseries, X–disease can be transmitted by grafting as well as by leafhoppers. Budwood trees should be examined periodically for symptoms of X–disease and infected trees removed.

Selected References

Gilmer, R. M., D. H. Palmiter, G. A. Schaffers and F. L. McEwen. 1966. *Insect transmission of X-disease virus of stone fruits in New York.* Bull. 813. Geneva, N.Y.: New York State Agricultural Experiment Station.

Jones, A. L., G. R. Hooper and D. A. Rosenberger. 1974. Association of mycoplasma-like bodies with little peach and X-disease. *Phytopathology,* 64:755–756.

Larsen, K. L., and M. E. Whalon. 1988. Crepuscular movement of *Paraphlepsius irroratus* (Say) (Homoptera:Cicadellidae) between the groundcover and cherry trees. *Environ. Entomol.,* 16:1103–1106.

Rosenberger, D. A., and A. L. Jones. 1978. Leafhopper vectors of peach X-disease and seasonal transmission from chokecherry. *Phytopathology,* 68:782–790.

Rosenberger, D. A., and A. L. Jones. 1977. Symptom remission in X-diseased peach trees as affected by date, method and rate of application of oxytetracycline-HCl. *Phytopathology,* 67:277–282.

Sinha, R. C., and L. N. Chiykowski. 1980. Transmission and morphological features of mycoplasma-like bodies associated with peach X-disease. *Can. J. Plant Pathol.,* 2:119–124.

Photo 158. Sweet cherry leaf with enlarged stipules due to X-disease.
(Courtesy S. V. Thomson, Utah State University, Logan)

Photo 159. *Paraphlepsius irroratus,* one of several species of leafhoppers that can spread X-disease.

Stone Fruits

PRUNUS STEM PITTING/ PRUNE BROWNLINE

Prunus stem pitting, also called prune brownline and constriction disease, is an important disease of stone fruits throughout the mid-Atlantic states and in isolated areas of the northeastern United States. In Michigan, it is a significant problem in the production of Stanley prunes.

Tomato ringspot virus (TmRSV), a nematode-transmitted virus, causes the disease. TmRSV is also a problem on apple (see "Union Necrosis and Decline" under Pome Fruit Diseases).

Symptoms

Affected peach and cherry trees look stunted and unthrifty (*Photo 160*). Their root systems are poorly developed, and when trees are pulled out of the ground, socket-like depressions remain where roots have broken away. These trees may break off slightly below ground level in high winds, exposing the disorganized wood in cross-section.

Foliar symptoms develop in late summer after harvest but before the time of normal defoliation. Leaves on affected trees cup upward, become yellowish, later turn reddish to purplish and fall to the ground prematurely. These symptoms indicate a root or vascular dysfunction and are not diagnostic.

When the bark is removed from the wood below ground level, elongated pits or indentations are evident (*Photo 161*). The severity of pitting depends on the cultivar or rootstock clone and the stage of disease development. In addition, the bark is often 1 to 2 inches thicker than normal and has a spongy texture.

Affected Stanley prune trees exhibit smaller than normal, pale green to yellow leaves, reduced tree growth and suckering from the Myrobalan rootstock. When the soil is removed from around the trunk of unthrifty trees, the shank of the Myrobalan rootstock is often restricted on one side and is smaller in diameter than the scion tissue immediately above the union (*Photo 162*). When bark is removed at the graft union, a dark brown line is often observed in the wood and phloem tissues (*Photo 163*). Affected trees decline for 3 to 5 years before dying.

Disease Cycle

TmRSV can be introduced into orchard plantings from infected nursery stock or seeds of infected

Photo 160. Peach tree in decline and healthy tree of the same age.

Photo 161. Sour cherry tree with bark removed to show pitting in the Mahaleb rootstock caused by Prunus stem pitting.

Photo 162. Prunus stem pitting on Stanley plum. Note constriction of Myrobalan rootstock and shoots from the trunk that indicate the constriction.
(Courtesy [left] S. N. Jeffers, Clemson University, Clemson, S.C.)

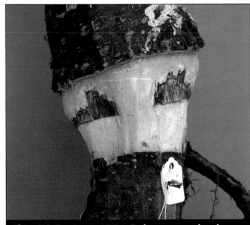

Photo 163. Prunus stem pitting. Bark was removed to show a wide line of dead tissue at the union between the Stanley plum scion and the Myrobalan rootstock.
(Courtesy D. A. Rosenberger, Hudson Valley Laboratory, Cornell University, Highland, N.Y.)

weeds. Once the virus is established in the orchard, it is spread by dagger nematodes and by seeds of infected weeds. The virus can persist in orchards for many years in common weeds such as dandelion and in nematode vectors.

Control

To avoid introducing this disease into new plantings, purchase certified virus-free trees grown before propagation on soil fumigated to control nematodes.

If replanting of infested orchard sites is attempted, the nematode vectors and weeds can be reduced by cultivating the site for 2 years before establishing a new orchard. Soil fumigation can be used to reduce nematodes and weeds, but it is not economical nor environmentally desirable.

Future control of TmRSV may be based on resistance derived from the genome of the virus itself. High-level resistance to TmRSV has been obtained by placing a modified TmRSV coat protein gene into tobacco plants. The next step is to build this resistance into important fruit crops.

Selected References

Cummins, J. N., and D. Gonsalves. 1986. Constriction and decline of Stanley prune associated with tomato ringspot virus. *J. Am. Soc. Hortic. Sci.,* 11:315–318.

Forer, L. B., C. A. Powell and R. F. Stouffer. 1984. Transmission of tomato ringspot virus to apple rootstock cuttings and to cherry and peach seedlings by *Xiphinema rivesi. Plant Dis.,* 68:1052–1054.

Mircetich, S. M., H. W. Fogle and E. L. Civerolo. 1970. Peach stem pitting: Transmission and natural spread. *Phytopathology,* 60:1329–1334.

Powell, C. A., L. B. Forer, R. F. Stouffer, J. N. Cummins, D. Gonsalves, D. A. Rosenberger, J. Hoffmann and R. M. Lister. 1984. Orchard weeds as hosts of tomato ringspot and tobacco ringspot viruses. *Plant Dis.,* 68:242–244.

Rosenberger, D. A., and F. W. Meyer. 1988. Control of dagger and lesion nematodes in apple and plum orchards with fenamiphos, carbofuran and carbosulfan. *Plant Dis.,* 72:519–522.

Smith, S. H., R. F. Stouffer and D. M. Soulen. 1973. Induction of stem pitting in peaches by mechanical inoculation with tomato ringspot virus. *Phytopathology,* 63:1404–1406.

Yepes, L. M., M. Fuchs, J. L. Slightom and D. Gonsalves. 1996. Sense and antisense coat protein constructs confer high levels of resistance to tomato ringspot nepovirus in transgenic *Nicotiana* species. *Phytopathology,* 86:417–424.

Stone Fruits

PEACH ROSETTE MOSAIC

Peach rosette mosaic, caused by peach rosette mosaic virus (PRMV), is a disease of peach in southwestern Michigan and in southwestern Ontario, Canada. It also causes a decline of Concord grapevines.

Symptoms and Disease Cycle

Infected trees exhibit a few twigs or limbs with darker than normal but small leaves that are crowded together (*Photo 164*). The crowding is due to the shortening of internodes on affected limbs and is referred to as rosetting. Symptoms will eventually develop throughout most of the tree.

PRMV is a nematode-transmitted, soilborne virus. It is transmitted by dagger (*Xiphinema* spp.) and needle (*Longidorus* spp.) nematodes from diseased peach trees to healthy trees. Natural field spread is slow. The virus moves systemically to aboveground portions of the tree. Trees infected when young may exhibit symptoms throughout the tree, while trees infected when mature may exhibit symptoms on isolated limbs.

Control

Although PRMV is graft transmissible, there has been little spread of the virus by infected scions, possible because the striking symptoms and limited natural distribution of the disease in peach make it easy to avoid.

Photo 164. Peach tree infected when young with peach rosette mosaic. Note compact growth due to rosetting.

Selected References

Allen, W. R., J. G. Van Schagen and B. A. Ebsary. 1984. Comparative transmission of the peach rosette mosaic virus by Ontario populations of *Longidorus diadecturus* and *Xiphinema americanum* (Nematoda: Longidoridae). *Can. J. Plant Pathol.,* 6:29–32.

Allen W. R., J. G. Van Schagen and E. S. Eveleigh. 1982. Transmission of peach rosette mosaic virus to peach, grape and cucumber by *Longidorus diadecturus* obtained from diseased orchards in Ontario. *Can. J. Plant Pathol.,* 4:16–18.

Dias, H. F., and D. Cation. 1976. The characterization of a virus responsible for peach rosette mosaic and grape decline in Michigan. *Can. J. Bot.,* 54:1228–1239.

PRUNUS NECROTIC RINGSPOT

Prunus necrotic ringspot–also called necrotic leafspot, rugose mosaic, shothole, Stecklingberger, tatter leaf and lace leaf–is caused by Prunus necrotic ringspot virus (PNRSV). It occurs on sweet and sour cherry, peach, nectarine, plum and prune.

PNRSV can cause considerable economic loss, depending on virus strain, fruit species, cultivar and climate. In years when symptoms of Prunus necrotic ringspot are acute, yields of sour cherry are reduced markedly. Losses for individual trees range from 25 to 50 percent.

Symptoms

Symptoms of Prunus necrotic ringspot in the first year after infection are severe. This stage of the disease is called the acute or "shock" stage. This is when greatest yield reductions occur.

There may be delayed bud break, death of leaf and flower buds, terminal dieback and gumming. Leaves on affected limbs are small, often with depressed partial to complete chlorotic or necrotic rings and arcs on the upper surfaces (Photo 165). The centers of the necrotic rings often fall out, giving the leaves a tattered appearance (Photo 166). Leaf symptoms are most pronounced during the 2-week period after petal fall. Symptoms are mild or absent on leaves that unfold later in the season. Green fruit only rarely show arcs and rings similar to those on leaves. Over a period of years, other limbs on the same tree will show acute symptoms, but, except for less common recurrent strains, symptoms rarely reappear on limbs that exhibited symptoms previously. Later, infected trees appear normal, though tree vigor and size may be reduced.

Disease Cycle

In the nursery, PNRSV is seedborne and can be transmitted by grafting. Therefore, PNRSV can be introduced into new orchards through infected nursery stock.

Movement of PNRSV into new plantings established with healthy trees occurs via virus-contaminated pollen from other orchards. New orchards often remain free of Prunus necrotic ringspot until they begin to bloom heavily. Once the virus is established in an orchard, it spreads from tree to tree with pollen. The rate of spread and buildup in a new planting is related to proximity to older, infected trees.

Once transmission occurs, a few weeks to 2 years are required before symptoms develop. The severity of the symptoms depends on temperature. Leaf symptoms develop readily between 68 and 75 degrees F but may occur between 50 and 82 degrees F. Dieback of shoots

Photo 165. Depressed, concentric rings on the upper surface of a sour cherry leaf due to prunus necrotic ringspot virus.

Photo 166. Pattern of shock symptoms on spur leaves of sour cherry caused by Prunus necrotic ringspot virus.

and spurs is more severe at
higher temperatures.

Control

As a result of nursery certification programs, new trees are
virtually free of the disease. In
addition, see control procedures
described for sour cherry yellows.

Selected References

Nyland, G., R. M. Gilmer and J. D. Moore. 1976. "Prunus" ring spot group.
Pages 104–132 in: *Virus Diseases and Noninfectious Disorders of Stone
Fruits in North America*. Agr. Handbook 437. Washington, D.C.: U.S.
Department of Agriculture.

Fulton, R. W. 1959. Purification of sour cherry necrotic ringspot and prune
dwarf viruses. *Virology*, 9:522–535.

Scott, S. W., O. W. Barnett and P. M. Burrows. 1989. Incidence of Prunus
necrotic ringspot virus in selected peach orchards in South Carolina.
Plant Dis., 73:913–916.

Wells, J. M., and H. C. Kirkpatrick. 1986. Symptomatology and incidence of
Prunus necrotic ringspot virus in peach orchards in Georgia. *Plant Dis.*,
70:444–447.

SOUR CHERRY YELLOWS

Sour cherry yellows–also called shot–hole, narrow leaf, chlorotic ringspot, blind wood and tatter leaf–is caused by the prune dwarf virus (PDV). It is the most serious disease of Montmorency sour cherry in the Great Lakes fruit belt.

Initial losses from PDV are not spectacular as with Prunus necrotic ringspot virus (PNRSV), but tree vigor and productivity decline with time. Yield potentials of sour cherry trees infected while young are reduced by 40 to 50 percent. The longer a new planting can be kept virus–free, the closer it will come to reaching its maximum yield potential.

Symptoms

Sour cherry yellows derives its name from the characteristic yellowing of the leaves (*Photo 167*). Symptoms are most severe on trees infected with PNRSV and begin to appear 2 to 3 years after the shock symptoms associated with Prunus necrotic ringspot. Some leaves drop without first turning yellow, but most show various degrees of yellow–green mottling. Defoliation generally occurs in waves beginning 3 to 4 weeks after petal fall. The severity of each wave depends on the temperature 30 days prior to the period of defoliation–low temperatures increase later symptom development.

PDV–infected trees develop an excess of flower buds on terminals and lateral shoots. As trees age, the bearing surface is reduced because no lateral vegetative buds remain to produce new fruit spurs. Severely infected trees exhibit willowy growth with long, bare spaces lacking fruit or spurs.

Flower buds on PDV–infected trees are weaker and more susceptible than healthy buds to low-temperature damage. Bloom is often staggered, leading to uneven fruit ripening. Fruit produced on old PDV–infected trees are larger than normal and, because of their size, may be objectionable for processing. Sweet cherry trees are also infected by PDV, but symptoms are rarely produced.

Disease Cycle

Because PDV is seed-borne, pollenborne and graft–transmitted, disease spread in the nursery and the orchard is as described for Prunus necrotic ringspot.

Control

Once a tree becomes infected, it will remain infected because there are no orchard treatments that can be used to cure infected trees. The following program has been shown to retard the movement of PNRSV and PDV into cherry plantings. Failure to follow any one practice will defeat the objectives of the program.

Purchase virus-free trees only. Modern nurseries obtain seeds for rootstocks from indexed trees and use virus–free buds to produce trees as free as possible from known viruses. Most states have inspection programs and certify trees indexed free of known viruses. Trees not certified free of viruses should be refused.

Isolate plantings. New orchards should be planted at least 100 feet (preferably 500 feet or more) from existing cherry blocks. Spread of PNRSV and PDV is rapid when older, infected trees are nearby. One way to isolate plantings is to establish apple or pear orchards between cherry orchards.

Plant solid blocks. Replanting missing trees is profitable only for the first 5 years. Thereafter, avoid replanting. When an orchard becomes unprofitable, remove it completely.

Selected References

Gilmer, R. M., G. Nyland and J. D. Moore. 1976. Prune dwarf. Pages 179–190 in: *Virus Diseases and Noninfectious Disorders of Stone Fruits in North America.* Agr. Handbook 437. Washington, D.C.: U.S. Department of Agriculture.

Gilmer, R. M., and R. D. Way. 1960. Pollen transmission of necrotic ringspot and prune dwarf virus in sour cherry. *Phytopathology,* 50:624–625.

Photo 167. Yellowing and defoliation of sour cherry leaves due to sour cherry yellows.

Stone Fruits

GREEN RING MOTTLE

Green ring mottle is found primarily in sour cherry orchards established with trees that were not virus certified. It also occurs without symptoms in peach, apricot and sweet cherry.

The recent sequencing of the green ring mottle virus (GRMV) genome, which shares extensive sequence homology with apple stem pitting virus and a few other viruses, should result in improved detection methods for this virus.

Symptoms and Disease Cycle

Infected sour cherry trees develop yellow leaves similar to those on trees with sour cherry yellows except for the presence of prominent dark green blotches or rings (Photo 168). Leaf symptoms often occur in waves extending from late June to mid-July, sometimes later. Defoliation is generally not as severe from green ring mottle as from cherry yellows. Both diseases may be present in the same tree and show symptoms, either concurrently or separately. A smaller number of leaves may show yellowing along a lateral vein and exhibit slight leaf distortion at the tip of the vein. This symptom is referred to as "constricting chlorosis" (Photo 169).

Fruit on Montmorency trees with green ring mottle may be indented with streaks of dead tissue extending to the pit. The distorted fruit are often worthless.

The disease is spread by grafting with infected propagating materials. Weed hosts are known, but their role in spread is not known. The virus is not transmitted by seed.

Control

Virus certification programs are effective for control. Establish new sour cherry orchards with certified trees.

Selected References

Barksdale, T. H. 1959. Green ring mottle virus as an entity distinct from sour cherry ring spot and yellows viruses. *Phytopathology*, 49:777–784.

Parker, K. G., P. R. Fridlund and R. M. Gilmer. 1976. Green ring mottle. Pages 193–199 in: *Virus Diseases and Noninfectious Disorders of Stone Fruits in North America.* Agr. Handbook 437. Washington, D.C.: U.S. Department of Agriculture.

Zagula, K. R., N. M. Aref and D. C. Ramsdell. 1989. Purification, serology and some properties of a mechanically transmissible virus associated with green ring mottle disease in peach and cherry. *Phytopathology*, 79:451–456.

Zhang, Y.-P., J. K. Uyemoto and B. C. Kirkpatrick. 1996. Nucleotide sequence and molecular characterization of sour cherry green ring mottle virus. (Abstr.) *Phytopathology*, 86:(in press).

Photo 168. Dark green arcs and rings on yellow background in Montmorency cherry leaves are typical of green ring mottle.

Photo 169. Constricting chlorosis symptoms on Montmorency cherry leaves caused by green ring mottle.

Diseases of Tree Fruits in the East
NCR 45

North Central Regional Extension resources are subject to peer review and prepared as a part of extension activities of the 13 land-grant universities in 12 North Central states, in cooperation with the Cooperative State Research, Education and Extension Service (CSREES)—U.S. Department of Agriculture, Washington, D.C.

For copies of this and other North Central Regional Extension resources, contact the distribution office of the university listed below for your state. If your university is not listed, contact the producing university (marked with an asterisk: ✳)

Iowa State University
Extension Distribution Center
119 Printing & Pub. Bldg.
Ames, IA 50011-3171
(515) 294-5247

Kansas State University
Distribution Center
Umberger Hall
Manhattan, KS 66506-3400
(913) 532-5830

✳Michigan State University
Bulletin Office
10-B Agriculture Hall
East Lansing, MI 48824-1039
(517) 355-0240

University of Minnesota
Distribution Center
20 Coffey Hall, 1420 Eckles Ave.
St. Paul, MN 55108-6064
(612) 625-8173

North Dakota State Univ.
Extension Communications
Box 5655, Morrill Hall
Fargo, ND 58105
(701) 231-7882

Ohio State University
Publications Office
385 Kottman Hall
2021 Coffey Rd.
Columbus, OH 43210-1044
(614) 292-9607

South Dakota State Univ.
Ag. Comm. Center
Box 2231
Brookings, SD 57007-0892
(605) 688-5628

University of Wisconsin
Cooperative Extension Publications
Rm. 245
30 N. Murray St.
Madison, WI 53715-2609
(608) 262-3346

In cooperation with IDEA (Information Development for Extension Audiences).

Issued in furtherance of Cooperative Extension work, Acts of Congress of May 8 and June 30, 1914, in cooperation with the U.S. Department of Agriculture and Cooperative Extension Services of Illinois, Indiana, Iowa, Kansas, Michigan, Minnesota, Missouri, Nebraska, North Dakota, Ohio, South Dakota and Wisconsin. Arlen Leholm, interim director, MSU Extension, East Lansing, MI 48824.

November 1996

✳*Publishing University*